Universitext

Springer
Berlin
Heidelberg
New York
Hong Kong
London
Milan
Paris
Tokyo

Jean Jacod
Philip Protter

Probability Essentials

Second Edition

 Springer

Jean Jacod
Université de Paris
Laboratoire de Probabilités
4, place Jussieu - Tour 56
75252 Paris Cedex 05, France

jj@ccr.jussieu.fr

Philip Protter
School of Operations Research
and Industrial Engineering
Cornell University
219 Rhodes Hall, Ithaca, NY 14853, USA

pep4@cornell.edu

Corrected Second Printing 2004

Cover art:
· Photograph of Paul Lévy by kind permission of Jean-Claude Lévy, Denise Piron,
  and Marie-Hélène Schwartz.
· Sketch of Carl Friedrich Gauß (by J.B. Listing; Nachlass Gauß, Posth. 26)
  by kind permission of Universitätsbibliothek Göttingen.
· Photograph of Andrei N. Kolmogorov by kind permission of Albert N. Shiryaev.

Library of Congress Cataloging-in-Publication Data applied for.

Bibliographic information published by *Die Deutsche Bibliothek*

Die Deutsche Bibliothek lists this publication in the Deutsche Nationalbibliografie;
detailed bibliographic data is available in the Internet at <http://dnb.ddb.de>.

ISBN 3-540-43871-8 2nd edition Springer-Verlag Berlin Heidelberg New York
ISBN 3-540-66419-X 1st edition Springer-Verlag Berlin Heidelberg New York

Springer-Verlag is a part of Springer Science+Business Media
springeronline.com

© Springer-Verlag Berlin Heidelberg 2000, 2003, 2004
Printed in Germany

Cover Design: Erich Kirchner, Springer-Verlag
Typesetting: Satztechnik Steingräber, Heidelberg from author´s data
Printing: Strauss Offsetdruck, Mörlenbach
Binding: J. Schäffer, Grünstadt

Printed on acid-free paper   41/3180PS 5 4 3 2 SPIN 12083640

To Diane and Sylvie
and
to Rachel, Margot, Olivier, Serge,
Thomas, Vincent and Martin

Preface to the Second Edition

We have made small changes throughout the book, including the exercises, and we have tried to correct if not all, then at least most of the typos. We wish to thank the many colleagues and students who have commented constructively on the book since its publication two years ago, and in particular Professors Valentin Petrov, Esko Valkeila, Volker Priebe, and Frank Knight.

Jean Jacod, Paris
Philip Protter, Ithaca
March, 2002

Preface to the Second Printing of the Second Edition

We have benefited greatly from the long list of typos and small suggestions sent to us by Professor Luis Tenorio. These corrections have improved the book in subtle yet important ways, and the authors are most grateful to him.

Jean Jacod, Paris
Philip Protter, Ithaca
January, 2004

Preface to the First Edition

We present here a one semester course on Probability Theory. We also treat measure theory and Lebesgue integration, concentrating on those aspects which are especially germane to the study of Probability Theory. The book is intended to fill a current need: there are mathematically sophisticated students and researchers (especially in Engineering, Economics, and Statistics) who need a proper grounding in Probability in order to pursue their primary interests. Many Probability texts available today are celebrations of Probability Theory, containing treatments of fascinating topics to be sure, but nevertheless they make it difficult to construct a lean one semester course that covers (what we believe) are the essential topics.

Chapters 1–23 provide such a course. We have indulged ourselves a bit by including Chapters 24–28 which are highly optional, but which may prove useful to Economists and Electrical Engineers.

This book had its origins in a course the second author gave in Perugia, Italy in 1997; he used the samizdat "notes" of the first author, long used for courses at the University of Paris VI, augmenting them as needed. The result has been further tested at courses given at Purdue University. We thank the indulgence and patience of the students both in Perugia and in West Lafayette. We also thank our editor Catriona Byrne, as well as Nick Bingham for many superb suggestions, an anonymous referee for the same, and Judy Mitchell for her extraordinary typing skills.

Jean Jacod, Paris
Philip Protter, West Lafayette

Table of Contents

1 Introduction

Almost everyone these days is familiar with the concept of Probability. Each day we are told the probability that it will rain the next day; frequently we discuss the probabilities of winning a lottery or surviving the crash of an airplane. The insurance industry calculates (for example) the probability that a man or woman will live past his or her eightieth birthday, given he or she is 22 years old and applying for life insurance. Probability is used in business too: for example, when deciding to build a waiting area in a restaurant, one wants to calculate the probability of needing space for more than n people each day; a bank wants to calculate the probability a loan will be repaid; a manufacturer wants to calculate the probable demand for his product in the future. In medicine a doctor needs to calculate the probability of success of various alternative remedies; drug companies calculate the probability of harmful side effects of drugs. An example that has recently achieved spectacular success is the use of Probability in Economics, and in particular in Stochastic Finance Theory. Here interest rates and security prices (such as stocks, bonds, currency exchanges) are modelled as varying randomly over time but subject to specific probability laws; one is then able to provide insurance products (for example) to investors by using these models. One could go on with such a list. Probability theory is ubiquitous in modern society and in science.

Probability theory is a reasonably old subject. Published references on games of chance (i.e., gambling) date to J. Cardan (1501–1576) with his book *De Ludo Alae* [4]. Probability also appears in the work of Kepler (1571–1630) and of Galileo (1564–1642). However historians seem to agree that the subject really began with the work of Pascal (1623–1662) and of Fermat (1601–1665). The two exchanged letters solving gambling "paradoxes" posed to them by the aristocrat de Méré. Later the Dutch mathematician Christian Huygens (1629–1695) wrote an influential book [13] elaborating on the ideas of Pascal and Fermat. Finally in 1685 it was Jacques Bernoulli (1654–1705) who proposed such interesting probability problems (in the "Journal des Scavans") (see also [3]) that it was necessary to develop a serious theory to answer them. After the work of J. Bernoulli and his contemporary A. De Moivre (1667–1754) [6], many renowned mathematicians of the day worked on probability problems, including Daniel Bernoulli (1700–1782), Euler (1707–1803),

Gauss (1777–1855), and Laplace (1749–1827). For a nice history of Probability before 1827 (the year of the death of Laplace) one can consult [21]. In the twentieth century it was Kolmogorov (1903–1987) who saw the connection between the ideas of Borel and Lebesgue and probability theory and he gave probability theory its rigorous measure theory basis. After the fundamental work of Kolmogorov, the French mathematician Paul Lévy (1886–1971) set the tone for modern Probability with his seminal work on Stochastic Processes as well as characteristic functions and limit theorems.

We think of Probability Theory as a mathematical model of chance, or random events. The idea is to start with a few basic principles about how the laws of chance behave. These should be sufficiently simple that one can believe them readily to correspond to nature. Once these few principles are accepted, we then deduce a mathematical theory to guide us in more complicated situations. This is the goal of this book.

We now describe the approach of this book. First we cover the bare essentials of discrete probability in order to establish the basic ideas concerning probability measures and conditional probability. We next consider probabilities on countable spaces, where it is easy and intuitive to fix the ideas. We then extend the ideas to general measures and of course probability measures on the real numbers. This represents Chapters 2–7. Random variables are handled analogously: first on countable spaces and then in general. Integration is established as the expectation of random variables, and later the connection to Lebesgue integration is clarified. This brings us through Chapter 12.

Chapters 13 through 21 are devoted to the study of limit theorems, the central feature of classical probability and statistics. We give a detailed treatment of Gaussian random variables and transformations of random variables, as well as weak convergence.

Conditional expectation is not presented via the Radon-Nikodym theorem and the Hahn–Jordan decomposition, but rather we use Hilbert Space projections. This allows a rapid approach to the theory. To this end we cover the necessities of Hilbert space theory in Chapter 22; we nevertheless extend the concept of conditional expectation beyond the Hilbert space setting to include integrable random variables. This is done in Chapter 23. Last, in Chapters 24–28 we give a beginning taste of martingales, with an application to the Radon–Nikodym Theorem. These last five chapters are not really needed for a course on the "essentials of probability". We include them however because many sophisticated applications of probability use martingales; also martingales serve as a nice introduction to the subject of stochastic processes.

We have written the book independent of the exercises. That is, the important material is in the text itself and not in the exercises. The exercises provide an opportunity to absorb the material by working with the subject. Starred exercises are suspected to be harder than the others.

We wish to acknowledge that Allan Gut's book [11] was useful in providing exercises, and part of our treatment of martingales was influenced by the delightful introduction to the book of Richard Bass [1].

No probability background is assumed. The reader should have a good knowledge of (advanced) calculus, some linear algebra, and also "mathematical sophistication".

Random Experiments

Random experiments are experiments whose output cannot be surely predicted in advance. But when one repeats the same experiment a large number of times one can observe some "regularity" in the average output. A typical example is the toss of a coin: one cannot predict the result of a single toss, but if we toss the coin many times we get an average of about 50% of "heads" if the coin is fair.

The theory of probability aims towards a mathematical theory which describes such phenomena. This theory contains three main ingredients:

a) The state space: this is the set of all possible outcomes of the experiment, and it is usually denoted by Ω.

Examples:

1) A toss of a coin: $\Omega = \{\text{h,t}\}$.
2) Two successive tosses of a coin: $\Omega = \{\text{hh,tt,ht,th}\}$.
3) A toss of two dice: $\Omega = \{(i,j) : 1 \le i \le 6, 1 \le j \le 6\}$.
4) The measurement of a length L, with a measurement error: $\Omega = \mathbf{R}_+$, where $\mathbf{R}_+$ denotes the positive real numbers $[0, \infty)$; $\omega \in \Omega$ denotes the result of the measurement, and $\omega - L$ is the measurement error.
5) The lifetime of a light-bulb: $\Omega = \mathbf{R}_+$.

b) The events: An "event" is a property which can be observed either to hold or not to hold *after* the experiment is done. In mathematical terms, an event is a subset of Ω. If A and B are two events, then

- the *contrary* event is interpreted as the complement set A^c;
- the event "A **or** B" is interpreted as the union $A \cup B$;
- the event "A **and** B" is interpreted as the intersection $A \cap B$;
- the *sure* event is Ω;
- the *impossible* event is the empty set $\emptyset$;
- an **elementary event** is a "singleton", i.e. a subset $\{\omega\}$ containing a single outcome ω of Ω.

We denote by $\mathcal{A}$ the family of all events. Often (but not always: we will see why later) we have $\mathcal{A} = 2^\Omega$, the set of all subsets of Ω. The family $\mathcal{A}$ should be "stable" by the logical operations described above: if $A, B \in \mathcal{A}$, then we must have $A^c \in \mathcal{A}$, $A \cap B \in \mathcal{A}$, $A \cup B \in \mathcal{A}$, and also $\Omega \in \mathcal{A}$ and $\emptyset \in \mathcal{A}$.

c) The probability: With each event A one associates a number denoted by $P(A)$ and called the "probability of A". This number measures the likelihood of the event A to be realized *a priori*, before performing the experiment. It is chosen between 0 and 1, and the more likely the event is, the closer to 1 this number is.

To get an idea of the properties of these numbers, one can imagine that they are the limits of the "frequency" with which the events are realized: let us repeat the same experiment n times; the n outcomes might of course be different (think of n successive tosses of the same die, for instance). Denote by $f_n(A)$ the frequency with which the event A is realized (i.e. the number of times the event occurs, divided by n). Intuitively we have:

$$P(A) \;=\; \text{limit of } f_n(A) \quad \text{as } n \uparrow +\infty. \tag{1.1}$$

(we will give a precise meaning to this "limit" later). From the obvious properties of frequencies, we immediately deduce that:

1. $0 \leq P(A) \leq 1$,
2. $P(\Omega) = 1$,
3. $P(A \cup B) = P(A) + P(B)$ if $A \cap B = \emptyset$.

A mathematical model for our experiment is thus a triple $(\Omega, \mathcal{A}, P)$, consisting of the space Ω, the family $\mathcal{A}$ of all events, and the family of all $P(A)$ for $A \in \mathcal{A}$; hence we can consider that P is a map from $\mathcal{A}$ into $[0, 1]$, which satisfies at least the properties (2) and (3) above (plus in fact an additional property, more difficult to understand, and which is given in Definition 2.3 of the next Chapter).

A fourth notion, also important although less basic, is the following one:

d) Random variable: A random variable is a quantity which depends on the outcome of the experiment. In mathematical terms, this is a map from Ω into a space E, where often $E = \mathbf{R}$ or $E = \mathbf{R}^d$. **Warning:** this terminology, which is rooted in the history of Probability Theory going back 400 years, is quite unfortunate; a random "variable" is not a variable in the analytical sense, but a function !

Let X be such a random variable, mapping Ω into E. One can then "transport" the probabilistic structure onto the target space E, by setting

$$P^X(B) = P(X^{-1}(B)) \quad \text{for} \quad B \subset E,$$

where $X^{-1}(B)$ denotes the pre-image of B by X, i.e. the set of all $\omega \in \Omega$ such that $X(\omega) \in B$. This formula defines a new probability, denoted by P^X, but on the space E instead of Ω. This probability P^X is called the **law of the variable** X.

Example (toss of two dice): We have seen that $\Omega = \{(i,j) : 1 \leq i \leq 6, 1 \leq j \leq 6\}$, and it is natural to take here $\mathcal{A} = 2^{\Omega}$ and

$$P(A) = \frac{\#(A)}{36} \qquad \text{if } A \subset \Omega,$$

where $\#(A)$ denotes the number of points in A. One easily verifies the properties (1), (2), (3) above, and $P(\{\omega\}) = \frac{1}{36}$ for each singleton. The map $X : \Omega \to \mathbf{N}$ defined by $X(i,j) = i + j$ is the random variable "sum of the two dice", and its law is

$$P_X(B) = \frac{\text{number of pairs } (i,j) \text{ such that } i + j \in B}{36}$$

(for example $P^X(\{2\}) = P(\{1,2\}) = \frac{1}{36}$, $P^X(\{3\}) = \frac{2}{36}$, etc...).

We will formalize the concept of a probability space in Chapter 2, and random variables are introduced with the usual mathematical rigor in Chapters 5 and 8.

2 Axioms of Probability

We begin by presenting the minimal properties we will need to define a Probability measure. Hopefully the reader will convince himself (or herself) that the two axioms presented in Definition 2.3 are reasonable, especially in view of the frequency approach (1.1). From these two simple axioms flows the entire theory. In order to present these axioms, however, we need to introduce the concept of a σ-algebras.

Let Ω be an abstract space, that is with no special structure. Let 2^{Ω} denote all subsets of Ω, including the empty set denoted by $\emptyset$. With $\mathcal{A}$ being a subset of 2^{Ω}, we consider the following properties:

1. $\emptyset \in \mathcal{A}$ and $\Omega \in \mathcal{A}$;
2. If $A \in \mathcal{A}$ then $A^c \in \mathcal{A}$, where A^c denotes the complement of A;
3. $\mathcal{A}$ is closed under finite unions and finite intersections: that is, if $A_1, \ldots, A_n$ are all in $\mathcal{A}$, then $\cup_{i=1}^{n} A_i$ and $\cap_{i=1}^{n} A_i$ are in $\mathcal{A}$ as well (for this it is enough that $\mathcal{A}$ be stable by the union and the intersection of any *two* sets);
4. $\mathcal{A}$ is closed under countable unions and intersections: that is, if $A_1, A_2, A_3, \ldots$ is a countable sequence of events in $\mathcal{A}$, then $\cup_{i=1}^{\infty} A_i$ and $\cap_{i=1}^{\infty} A_i$ are both also in $\mathcal{A}$.

Definition 2.1. $\mathcal{A}$ *is an* algebra *if it satisfies* (1), (2) *and* (3) *above. It is a* σ-algebra, *(or a* σ-field*) if it satisfies* (1), (2), *and* (4) *above.*

Note that under (2), (1) can be replaced by either (1'): $\emptyset \in \mathcal{A}$ or by (1''): $\Omega \in \mathcal{A}$. Note also that (1)+(4) implies (3), hence any σ-algebra is an algebra (but there are algebras that are not σ-algebras: see Exercise 2.17).

Definition 2.2. *If* $\mathcal{C} \subset 2^{\Omega}$, *the* σ-algebra generated by $\mathcal{C}$, *and written* $\sigma(\mathcal{C})$, *is the smallest* σ-algebra containing $\mathcal{C}$. *(It always exists because* 2^{Ω} *is a* σ-algebra, *and the intersection of a family of* σ-algebras is again a σ-algebra: See Exercise 2.2.)*

Example:

(i) $\mathcal{A} = \{\emptyset, \Omega\}$ (the trivial σ-algebra).
(ii) A is a subset; then $\sigma(A) = \{\phi, A, A^c, \Omega\}$.

(iii) If $\Omega = \mathbf{R}$ (the Real numbers) (or more generally if Ω is a space with a topology, a case we treat in Chapter 8), the *Borel σ-algebra* is the σ-algebra generated by the open sets (or by the closed sets, which is equivalent).

Theorem 2.1. *The Borel σ-algebra of $\mathbf{R}$ is generated by intervals of the form $(-\infty, a]$, where $a \in \mathbf{Q}$ ($\mathbf{Q}$ = rationals).*

Proof. Let $\mathcal{C}$ denote all open intervals. Since every open set in $\mathbf{R}$ is the countable union of open intervals, we have $\sigma(\mathcal{C}) =$ the Borel σ-algebra of $\mathbf{R}$.

Let $\mathcal{D}$ denote all intervals of the form $(-\infty, a]$, where $a \in \mathbf{Q}$. Let $(a, b) \in \mathcal{C}$, and let $(a_n)_{n \geq 1}$ be a sequence of rationals decreasing to a and $(b_n)_{n \geq 1}$ be a sequence of rationals increasing strictly to b. Then

$$(a, b) = \cup_{n=1}^{\infty} (a_n, b_n]$$
$$= \cup_{n=1}^{\infty} \left((-\infty, b_n] \cap (-\infty, a_n]^c \right),$$

Therefore $\mathcal{C} \subset \sigma(\mathcal{D})$, whence $\sigma(\mathcal{C}) \subset \sigma(\mathcal{D})$. However since each element of $\mathcal{D}$ is a closed set, it is also a Borel set, and therefore $\sigma(\mathcal{D})$ is contained in the Borel sets $\mathcal{B}$. Thus we have

$$\mathcal{B} = \sigma(\mathcal{C}) \subset \sigma(\mathcal{D}) \subset \mathcal{B},$$

and hence $\sigma(\mathcal{D}) = \mathcal{B}$. □

On the state space Ω the family of all events will always be a σ-algebra $\mathcal{A}$: the axioms (1), (2) and (3) correspond to the "logical" operations described in Chapter 1, while Axiom (4) is necessary for mathematical reasons. The probability itself is described below:

Definition 2.3. *A probability measure defined on a σ-algebra $\mathcal{A}$ of Ω is a function $P : \mathcal{A} \to [0, 1]$ that satisfies:*

1. $P(\Omega) = 1$
2. *For every countable sequence $(A_n)_{n \geq 1}$ of elements of $\mathcal{A}$, pairwise disjoint (that is, $A_n \cap A_m = \emptyset$ whenever $n \neq m$), one has*

$$P \left(\cup_{n=1}^{\infty} A_n \right) = \sum_{n=1}^{\infty} P(A_n).$$

Axiom (2) above is called *countable additivity*; the number $P(A)$ is called the *probability* of the event A.

In Definition 2.3 one might imagine a more naïve condition than (2), namely:

$$A, B \in \mathcal{A}, \ A \cap B = \emptyset \quad \Rightarrow \quad P(A \cup B) = P(A) + P(B). \qquad (2.1)$$

This property is called *additivity* (or "finite additivity") and, by an elementary induction , it implies that for every *finite* $A_1, \ldots A_m$ of pairwise disjoint events $A_i \in \mathcal{A}$ we have

$$P\left(\cup_{n=1}^m A_n\right) = \sum_{n=1}^m P(A_n).$$

Theorem 2.2. *If P is a probability measure on $(\Omega, \mathcal{A})$, then:*

(i) *We have $P(\emptyset) = 0$.*
(ii) *P is additive.*

Proof. If in Axiom (2) we take $A_n = \emptyset$ for all n, we see that the number $a = P(\emptyset)$ is equal to an infinite sum of itself; since $0 \leq a \leq 1$, this is possible only if $a = 0$, and we have (i). For (ii) it suffices to apply Axiom (2) with $A_1 = A$ and $A_2 = B$ and $A_3 = A_4 = \ldots = \emptyset$, plus the fact that $P(\emptyset) = 0$, to obtain (2.1). $\qquad\square$

Conversely, countable additivity is *not* implied by additivity. In fact, in spite of its intuitive appeal, additivity is not enough to handle the mathematical problems of the theory, even in such a simple example as tossing a coin, as we shall see later.

The next theorem (Theorem 2.3) shows exactly what is extra when we assume countable additivity instead of just finite additivity. Before stating this theorem, and to see that the last four conditions in it are meaningful, let us mention the following immediate consequence of Definition 2.3:

$$A, C \in \mathcal{A}, \quad A \subset C \quad \Rightarrow \quad P(A) \leq P(C)$$

(take $B = A^c \cap C$, hence $A \cap B = \emptyset$ and $A \cup B = C$, and apply (2.1)).

Theorem 2.3. *Let $\mathcal{A}$ be a σ-algebra. Suppose that $P : \mathcal{A} \to [0, 1]$ satisfies (1) and is additive. Then the following are equivalent:*

(i) *Axiom (2) of Definition (2.3).*
(ii) *If $A_n \in \mathcal{A}$ and $A_n \downarrow \emptyset$, then $P(A_n) \downarrow 0$.*
(iii) *If $A_n \in \mathcal{A}$ and $A_n \downarrow A$, then $P(A_n) \downarrow P(A)$.*
(iv) *If $A_n \in \mathcal{A}$ and $A_n \uparrow \Omega$, then $P(A_n) \uparrow 1$.*
(v) *If $A_n \in \mathcal{A}$ and $A_n \uparrow A$, then $P(A_n) \uparrow P(A)$.*

Proof. The notation $A_n \downarrow A$ means that $A_{n+1} \subset A_n$, each n, and $\cap_{n=1}^\infty A_n = A$. The notation $A_n \uparrow A$ means that $A_n \subset A_{n+1}$ and $\cup_{n=1}^\infty A_n = A$.

Note that if $A_n \downarrow A$, then $A_n^c \uparrow A^c$, and by the finite additivity axiom $P(A_n^c) = 1 - P(A_n)$. Therefore (ii) is equivalent to (iv) and similarly (iii) is equivalent to (v). Moreover by choosing A to be Ω we have that (v) implies (iv).

Suppose now that we have (iv). Let $A_n \in \mathcal{A}$ with $A_n \uparrow A$. Set $B_n = A_n \cup A^c$. Then B_n increases to Ω, hence $P(B_n)$ increases to 1. Since $A_n \subset A$ we have $A_n \cap A^c = \emptyset$, whence $P(A_n \cup A^c) = P(A_n) + P(A^c)$. Thus

$$1 = \lim_{n \to \infty} P(B_n) = \lim_{n \to \infty} \{P(A_n) + P(A^c)\},$$

whence $\lim_{n \to \infty} P(A_n) = 1 - P(A^c) = P(A)$, and we have (v).

It remains to show that (i) is equivalent to (v). Suppose we have (v). Let $A_n \in \mathcal{A}$ be *pairwise disjoint:* that is, if $n \neq m$, then $A_n \cap A_m = \emptyset$. Define $B_n = \cup_{1 \leq p \leq n} A_p$ and $B = \cup_{n=1}^{\infty} A_n$. Then by the definition of a Probability Measure we have $P(B_n) = \sum_{p=1}^{n} P(A_p)$ which increases with n to $\sum_{n=1}^{\infty} P(A_n)$, and also $P(B_n)$ increases to $P(B)$ by (v). We deduce $\lim_{n \to \infty} P(B_n) = P(B)$ and we have

$$P(B) = P(\cup_{n=1}^{\infty} A_n) = \sum_{n=1}^{\infty} P(A_n)$$

and thus we have (i).

Finally assume we have (i), and we wish to establish (v). Let $A_n \in \mathcal{A}$, with A_n increasing to A. We construct a new sequence as follows:

$$B_1 = A_1,$$
$$B_2 = A_2 \setminus A_1 = A_2 \cap (A_1^c),$$
$$\vdots$$
$$B_n = A_n \setminus A_{n-1}.$$

Then $\cup_{n=1}^{\infty} B_n = A$ and the events $(B_n)_{n \geq 1}$ are pairwise disjoint. Therefore by (i) we have

$$P(A) = \lim_{n \to \infty} \sum_{p=1}^{n} P(B_p).$$

But also $\sum_{p=1}^{n} P(B_p) = P(A_n)$, whence we deduce $\lim_{n \to \infty} P(A_n) = P(A)$ and we have (v). $\qquad \square$

If $A \in 2^{\Omega}$, we define the *indicator function* by

$$1_A(\omega) = \begin{cases} 1 \text{ if } \omega \in A, \\ 0 \text{ if } \omega \notin A. \end{cases}$$

We often do not explicitly write the ω, and just write 1_A.

We can say that $A_n \in \mathcal{A}$ *converges* to A (we write $A_n \to A$) if $\lim_{n \to \infty} 1_{A_n}(\omega) = 1_A(\omega)$ for all $\omega \in \Omega$. Note that if the sequence A_n increases (resp. decreases) to A, then it also tends to A in the above sense.

Theorem 2.4. *Let P be a probability measure, and let A_n be a sequence of events in $\mathcal{A}$ which converges to A. Then $A \in \mathcal{A}$ and $\lim_{n \to \infty} P(A_n) = P(A)$.*

Proof. Let us define

$$\limsup_{n \to \infty} A_n = \cap_{n=1}^{\infty} \cup_{m \geq n} A_m,$$
$$\liminf_{n \to \infty} A_n = \cup_{n=1}^{\infty} \cap_{m \geq n} A_m.$$

Since $\mathcal{A}$ is a σ-algebra, we have $\limsup_{n\to\infty} A_n \in \mathcal{A}$ and $\liminf_{n\to\infty} A_n \in \mathcal{A}$ (see Exercise 2.4).

By hypothesis A_n converges to A, which means $\lim_{n\to\infty} 1_{A_n} = 1_A$, all ω. This is equivalent to saying that $A = \limsup_{n\to\infty} A_n = \liminf_{n\to\infty} A_n$. Therefore $A \in \mathcal{A}$.

Now let $B_n = \cap_{m\geq n} A_m$ and $C_n = \cup_{m\geq n} A_m$. Then B_n increases to A and C_n decreases to A, thus $\lim_{n\to\infty} P(B_n) = \lim_{n\to\infty} P(C_n) = P(A)$, by Theorem 2.3. However $B_n \subset A_n \subset C_n$, therefore $P(B_n) \leq P(A_n) \leq P(C_n)$, so $\lim_{n\to\infty} P(A_n) = P(A)$ as well. $\qquad\square$

Exercises for Chapter 2

2.1 Let Ω be a finite set. Show that the set of all subsets of Ω, 2^Ω, is also finite and that it is a σ-algebra.

2.2 Let $(\mathcal{G}_\alpha)_{\alpha \in A}$ be an arbitrary family of σ-algebras defined on an abstract space Ω. Show that $\mathcal{H} = \cap_{\alpha \in A} \mathcal{G}_\alpha$ is also a σ-algebra.

2.3 Let $(A_n)_{n \geq 1}$ be a sequence of sets. Show that (De Morgan's Laws)

a) $(\cup_{n=1}^\infty A_n)^c = \cap_{n=1}^\infty A_n^c$
b) $(\cap_{n=1}^\infty A_n)^c = \cup_{n=1}^\infty A_n^c$.

2.4 Let $\mathcal{A}$ be a σ-algebra and $(A_n)_{n \geq 1}$ a sequence of events in $\mathcal{A}$. Show that

$$\liminf_{n \to \infty} A_n \in \mathcal{A}; \quad \limsup_{n \to \infty} A_n \in \mathcal{A}; \quad \text{and} \quad \liminf_{n \to \infty} A_n \subset \limsup_{n \to \infty} A_n.$$

2.5 Let $(A_n)_{n \geq 1}$ be a sequence of sets. Show that

$$\limsup_{n \to \infty} 1_{A_n} - \liminf_{n \to \infty} 1_{A_n} = 1_{\{\limsup_n A_n \setminus \liminf_n A_n\}}$$

(where $A \setminus B = A \cap B^c$ whenever $B \subset A$).

2.6 Let $\mathcal{A}$ be a σ-algebra of subsets of Ω and let $B \in \mathcal{A}$. Show that $\mathcal{F} = \{A \cap B : A \in \mathcal{A}\}$ is a σ-algebra of subsets of B. Is it still true when B is a subset of Ω that does not belong to $\mathcal{A}$?

2.7 Let f be a function mapping Ω to another space E with a σ-algebra $\mathcal{E}$. Let $\mathcal{A} = \{A \subset \Omega$: there exists $B \in \mathcal{E}$ with $A = f^{-1}(B)\}$. Show that $\mathcal{A}$ is a σ-algebra on Ω.

2.8 Let $f : \mathbf{R} \to \mathbf{R}$ be a continuous function, and let $\mathcal{A} = \{A \subset \mathbf{R}$: there exists $B \in \mathcal{B}$ with $A = f^{-1}(B)\}$ where $\mathcal{B}$ are the Borel subsets of the range space $\mathbf{R}$. Show that $\mathcal{A} \subset \mathcal{B}$, the Borel subsets of the domain space $\mathbf{R}$.

For problems 2.9–2.15 we assume a fixed abstract space Ω, a σ-algebra $\mathcal{A}$, and a Probability P defined on $(\Omega, \mathcal{A})$. The sets A, B, A_i, etc... always belong to $\mathcal{A}$.

2.9 For $A, B \in \mathcal{A}$ with $A \cap B = \emptyset$, show $P(A \cup B) = P(A) + P(B)$.

2.10 For $A, B \in \mathcal{A}$, show $P(A \cup B) = P(A) + P(B) - P(A \cap B)$.

2.11 For $A \in \mathcal{A}$, show $P(A) = 1 - P(A^c)$.

2.12 For $A, B \in \mathcal{A}$, show $P(A \cap B^c) = P(A) - P(A \cap B)$.

2.13 Let $A_1, \ldots, A_n$ be given events. Show that

$$P\left(\cup_{i=1}^n A_i\right) = \sum_i P(A_i) - \sum_{i<j} P(A_i \cap A_j)$$

$$+ \sum_{i<j<k} P(A_i \cap A_j \cap A_k) - \ldots + (-1)^{n+1} P(A_1 \cap A_2 \cap \ldots \cap A_n)$$

where (for example) $\sum_{i<j}$ means to sum over all ordered pairs (i, j) with $i < j$.

2.14 Suppose $P(A) = \frac{3}{4}$ and $P(B) = \frac{1}{3}$. Show that always $\frac{1}{12} \leq P(A \cap B) \leq \frac{1}{3}$.

2.15 (Subadditivity) Let $A_i \in \mathcal{A}$ be a sequence of events. Show that

$$P\left(\cup_{i=1}^n A_i\right) \leq \sum_{i=1}^n P(A_i),$$

each n, and also

$$P\left(\cup_{i=1}^\infty A_i\right) \leq \sum_{i=1}^\infty P(A_i).$$

2.16 (Bonferroni Inequalities) Let $A_i \in \mathcal{A}$ be a sequence of events. Show that

a) $P\left(\cup_{i=1}^n A_i\right) \geq \sum_{i=1}^n P(A_i) - \sum_{i<j} P(A_i \cap A_j),$

b) $P\left(\cup_{i=1}^n A_i\right) \leq \sum_{i=1}^n P(A_i) - \sum_{i<j} P(A_i \cap A_j) + \sum_{i<j<k} P(A_i \cap A_j \cap A_k).$

2.17 Suppose that Ω is an infinite set (countable or not), and let $\mathcal{A}$ be the family of all subsets which are either finite or have a finite complement. Show that $\mathcal{A}$ is an algebra, but not a σ-algebra.

3 Conditional Probability and Independence

Let A and B be two events defined on a probability space. Let $f_n(A)$ denote the number of times A occurs divided by n. Intuitively, as n gets large, $f_n(A)$ should be close to $P(A)$. Informally, we should have $\lim_{n\to\infty} f_n(A) = P(A)$ (see Chapter 1).

Suppose now that we know the event B has occurred. Let $P(A|B)$ denote "the probability of A occurring given knowledge that B has occurred". What should $P(A|B)$ be? If we look at $f_n(A)$, it is silly to count the occurrences of $A \cap B^c$, since we know B has occurred. Therefore if we only count those occurrences of A where B also occurs, this is $nf_n(A \cap B)$. Now the number of trials is the number of occurrences of B (all other trials are discarded as impossible since B has occurred). Therefore the number of relevant trials is $nf_n(B)$. Consequently we should have

$$P(A|B) \approx \frac{nf_n(A \cap B)}{nf_n(B)} = \frac{f_n(A \cap B)}{f_n(B)},$$

and "taking limits in n" motivates Definition 3.2 which follows.

Next imagine that events A and B are "independent" in the sense that knowledge that B has occurred in no way changes one's probability that A will occur. Then one should have $P(A \mid B) = P(A)$; this implies

$$\frac{P(A \cap B)}{P(B)} = P(A), \text{ or } P(A \cap B) = P(A)P(B).$$

This motivates Definition 3.1 which follows; the definition is a little more complicated in order to handle finite collections of events.

Definition 3.1. (a) *Two events A and B are* independent *if $P(A \cap B) = P(A)P(B)$.*
(b) *A (possibly infinite) collection of events $(A_i)_{i \in I}$ is an* independent *collection if for every finite subset J of I, one has*

$$P\left(\cap_{i \in J} A_i\right) = \prod_{i \in J} P(A_i).$$

The collection $(A_i)_{i \in I}$ is often said to be mutually independent.

Warning: If events $(A_i)_{i \in I}$ are independent, they are pairwise independent, but the converse is false. ($(A_i)_{i \in I}$ are *pairwise independent* if A_i and A_j are independent for all i, j with $i \neq j$.)

Theorem 3.1. *If A and B are independent, so also are A and B^c, A^c and B, and A^c and B^c.*

Proof. For A and B^c,

$$P(A \cap B^c) = P(A) - P(A \cap B) = P(A) - P(A)P(B) = P(A)(1 - P(B))$$
$$= P(A)P(B^c).$$

The other implications have analogous proofs.   □

Examples:

1. Toss a coin 3 times. If A_i is an event depending only on the ith toss, then it is standard to model $(A_i)_{1 \leq i \leq 3}$ as being independent.
2. One chooses a card at random from a deck of 52 cards. $A = \{$the card is a heart$\}$, and $B = \{$the card is Queen$\}$. A natural model for this experiment consists in prescribing the probability $\frac{1}{52}$ for picking any one of the cards. By additivity, $P(A) = \frac{13}{52}$ and $P(B) = \frac{4}{52}$ and $P(A \cap B) = \frac{1}{52}$, hence A and B are independent.
3. Let $\Omega = \{1, 2, 3, 4\}$, and $\mathcal{A} = 2^{\Omega}$. Let $P(i) = \frac{1}{4}$, where $i = 1, 2, 3, 4$. Let $A = \{1, 2\}$, $B = \{1, 3\}$, $C = \{2, 3\}$. Then A, B, C are pairwise independent but are not independent.

Definition 3.2. *Let A, B be events, $P(B) > 0$. The conditional probability of A given B is $P(A \mid B) = P(A \cap B)/P(B)$.*

Theorem 3.2. *Suppose $P(B) > 0$.*

(a) *A and B are independent if and only if $P(A \mid B) = P(A)$.*
(b) *The operation $A \to P(A \mid B)$ from $\mathcal{A} \to [0, 1]$ defines a new probability measure on $\mathcal{A}$, called the "conditional probability measure given B".*

Proof. We have already established (a) in the discussion preceding the theorem. For (b), define $Q(A) = P(A \mid B)$, with B fixed. We must show Q satisfies (1) and (2) of Definition 2.3. But

$$Q(\Omega) = P(\Omega \mid B) = \frac{P(\Omega \cap B)}{P(B)} = \frac{P(B)}{P(B)} = 1.$$

Therefore, Q satisfies (1). As for (2), note that if $(A_n)_{n \geq 1}$ is a sequence of elements of $\mathcal{A}$ which are pairwise disjoint, then

$$Q\left(\cup_{n=1}^{\infty} A_n\right) = P\left(\cup_{n=1}^{\infty} A_n \mid B\right) = \frac{P((\cup_{n=1}^{\infty} A_n) \cap B)}{P(B)} = \frac{P(\cup_{n=1}^{\infty}(A_n \cap B))}{P(B)}$$

and also the sequence $(A_n \cap B)_{n \geq 1}$ is pairwise disjoint as well; thus

$$= \sum_{n=1}^{\infty} \frac{P(A_n \cap B)}{P(B)} = \sum_{n=1}^{\infty} P(A_n \mid B) = \sum_{n=1}^{\infty} Q(A_n).$$

$\square$

The next theorem connects independence with conditional probability for a finite number of events.

Theorem 3.3. *If $A_1, \ldots, A_n \in \mathcal{A}$ and if $P(A_1 \cap \ldots \cap A_{n-1}) > 0$, then*

$$P(A_1 \cap \ldots \cap A_n) = P(A_1)P(A_2 \mid A_1)P(A_3 \mid A_1 \cap A_2)$$
$$\ldots P(A_n \mid A_1 \cap \ldots \cap A_{n-1}).$$

Proof. We use induction. For $n = 2$, the theorem is simply Definition 3.2. Suppose the theorem holds for $n - 1$ events. Let $B = A_1 \cap \ldots \cap A_{n-1}$. Then by Definition 3.2 $P(B \cap A_n) = P(A_n \mid B)P(B)$; next we replace $P(B)$ by its value given in the inductive hypothesis:

$$P(B) = P(A_1)P(A_2 \mid A_1) \ldots P(A_{n-1} \mid A_1 \cap \ldots \cap A_{n-2}),$$

and we get the result. $\square$

A collection of events (E_n) is called a *partition* of Ω if $E_n \in \mathcal{A}$, each n, they are pairwise disjoint, $P(E_n) > 0$, each n, and $\cup_n E_n = \Omega$.

Theorem 3.4 (Partition Equation). *Let $(E_n)_{n \geq 1}$ be a finite or countable partition of Ω. Then if $A \in \mathcal{A}$,*

$$P(A) = \sum_n P(A \mid E_n)P(E_n).$$

Proof. Note that

$$A = A \cap \Omega = A \cap (\cup_n E_n) = \cup_n (A \cap E_n).$$

Since the E_n are pairwise disjoint so also are $(A \cap E_n)_{n \geq 1}$, hence

$$P(A) = P(\cup_n (A \cap E_n)) = \sum_n P(A \cap E_n) = \sum_n P(A \mid E_n)P(E_n).$$

$\square$

Theorem 3.5 (Bayes' Theorem). *Let (E_n) be a finite or countable partition of Ω, and suppose $P(A) > 0$. Then*

$$P(E_n \mid A) = \frac{P(A \mid E_n)P(E_n)}{\sum_m P(A \mid E_m)P(E_m)}.$$

Proof. By Theorem 3.4 we have that the denominator

$$\sum_n P(A \mid E_m)P(E_m) = P(A).$$

Therefore the formula becomes

$$\frac{P(A \mid E_n)P(E_n)}{P(A)} = \frac{P(A \cap E_n)}{P(A)} = P(E_n \mid A).$$

$\square$

Bayes' theorem is quite simple but it has profound consequences both in Probability and Statistics. See, for example, Exercise 3.6.

Exercises for Chapter 3

In all exercises the probability space is fixed, and A, B, A_n, etc... are events.

3.1 Show that if $A \cap B = \emptyset$, then A and B *cannot* be independent unless $P(A) = 0$ or $P(B) = 0$.

3.2 Let $P(C) > 0$. Show that $P(A \cup B \mid C) = P(A \mid C) + P(B \mid C) - P(A \cap B \mid C)$.

3.3 Suppose $P(C) > 0$ and $A_1, \ldots, A_n$ are all pairwise disjoint. Show that

$$P(\cup_{i=1}^n A_i \mid C) = \sum_{i=1}^n P(A_i \mid C).$$

3.4 Let $P(B) > 0$. Show that $P(A \cap B) = P(A \mid B)P(B)$.

3.5 Let $0 < P(B) < 1$ and A be any event. Show

$$P(A) = P(A \mid B)P(B) + P(A \mid B^c)P(B^c).$$

3.6 Donated blood is screened for AIDS. Suppose the test has 99% accuracy, and that one in ten thousand people in your age group are HIV positive. The test has a 5% false positive rating, as well. Suppose the test screens you as positive. What is the probability you have AIDS? Is it 99%? (*Hint:* 99% refers to P (test positive|you have AIDS). You want to find P (you have AIDS|test is positive).

3.7 Let $(A_n)_{n \geq 1} \in \mathcal{A}$ and $(B_n)_{n \geq 1} \in \mathcal{A}$ and $A_n \to A$ (see before Theorem 2.4 for the definition of $A_n \to A$) and $B_n \to B$, with $P(B) > 0$ and $P(B_n) > 0$, all n. Show that

a) $\lim_{n \to \infty} P(A_n \mid B) = P(A \mid B)$,
b) $\lim_{n \to \infty} P(A \mid B_n) = P(A \mid B)$.
c) $\lim_{n \to \infty} P(A_n \mid B_n) = P(A \mid B)$.

3.8 Suppose we model tossing a coin with two outcomes, H and T, representing Heads and Tails. Let $P(H) = P(T) = \frac{1}{2}$. Suppose now we toss two such coins, so that the sample space of outcomes Ω consists of four points: HH, HT, TH, TT. We assume that the tosses are independent.

a) Find the conditional probability that both coins show a head given that the first shows a head (answer: $\frac{1}{2}$).
b) Find the conditional probability that both coins show heads given that at least one of them is a head (answer: $\frac{1}{3}$).

3.9 Suppose A, B, C are independent events and $P(A \cap B) \neq 0$. Show $P(C \mid A \cap B) = P(C)$.

3.10 A box has r red and b black balls. A ball is chosen at random from the box (so that each ball is equally likely to be chosen), and then a second ball is drawn at random from the remaining balls in the box. Find the probabilities that

a) Both balls are red $\left[\text{Ans.:}\ \frac{r(r-1)}{(r+b)(r+b-1)}\right]$

b) The first ball is red and the second is black $\left[\text{Ans.:}\ \frac{rb}{(r+b)(r+b-1)}\right]$

3.11 (Polya's Urn) An urn contains r red balls and b blue balls. A ball is chosen at random from the urn, its color is noted, and it is returned together with d more balls of the same color. This is repeated indefinitely. What is the probability that

a) The second ball drawn is blue? $\left[\text{Ans.:}\ \frac{b}{b+r}\right]$

b) The first ball drawn is blue given that the second ball drawn is blue?
$\left[\text{Ans.:}\ \frac{b+d}{b+r+d}\right]$

3.12 Consider the framework of Exercise 3.11. Let B_n denote the event that the nth ball drawn is blue. Show that $P(B_n) = P(B_1)$ for all $n \geq 1$.

3.13 Consider the framework of Exercise 3.11. Find the probability that the first ball is blue given that the n subsequent drawn balls are all blue. Find the limit of this probability as n tends to ∞. $\left[\text{Ans.:}\ \frac{b+nd}{b+r+nd};\ \text{limit is } 1\right]$

3.14 An insurance company insures an equal number of male and female drivers. In any given year the probability that a male driver has an accident involving a claim is α, independently of other years. The analogous probability for females is β. Assume the insurance company selects a driver at random.

a) What is the probability the selected driver will make a claim this year?
$\left[\text{Ans.:}\ \frac{\alpha+\beta}{2}\right]$

b) What is the probability the selected driver makes a claim in two consecutive years? $\left[\text{Ans.:}\ \frac{\alpha^2+\beta^2}{2}\right]$

3.15 Consider the framework of Exercise 3.14 and let A_1, A_2 be the events that a randomly chosen driver makes a claim in each of the first and second years, respectively. Show that $P(A_2 \mid A_1) \geq P(A_1)$.
$\left[\text{Ans.:}\ P(A_2 \mid A_1) - P(A_1) = \frac{(\alpha-\beta)^2}{2(\alpha+\beta)}\right]$

3.16 Consider the framework of Exercise 3.14 and find the probability that a claimant is female. $\left[\text{Ans.:}\ \frac{\beta}{\alpha+\beta}\right]$

3.17 Let $A_1, A_2, \ldots, A_n$ be independent events. Show that the probability that none of the $A_1, \ldots, A_n$ occur is less than or equal to $\exp(-\sum_{i=1}^{n} P(A_i))$.

3.18 Let A, B be events with $P(A) > 0$. Show $P(A \cap B \mid A \cup B) \leq P(A \cap B \mid A)$.

4 Probabilities on a Finite or Countable Space

For Chapter 4, we assume Ω is finite or countable, and we take the σ-algebra $\mathcal{A} = 2^\Omega$ (the class of all subsets of Ω).

Theorem 4.1. (a) *A probability on the finite or countable set Ω is characterized by its values on the atoms:* $p_\omega = P(\{\omega\})$, $\omega \in \Omega$.

(b) *Let* $(p_\omega)_{\omega \in \Omega}$ *be a family of real numbers indexed by the finite or countable set Ω. Then there exists a unique probability P such that* $P(\{\omega\}) = p_\omega$ *if and only if* $p_\omega \geq 0$ *and* $\sum_{\omega \in \Omega} p_\omega = 1$.

When Ω is countably infinite, $\sum_\omega p_\omega$ is the sum of an infinite number of terms which *a priori* are not ordered: although it is possible to enumerate the points of Ω, such an enumeration is in fact arbitrary. So we do not have a proper series, but rather a "summable family". In the Appendix to this chapter we gather some useful facts on summable families.

Proof. Let $A \in \mathcal{A}$; then $A = \cup_{\omega \in A}\{\omega\}$, a finite or countable union of pairwise disjoint singletons. If P is a probability, countable additivity yields

$$P(A) = P\left(\cup_{\omega \in A}\{\omega\}\right) = \sum_{\omega \in A} P(\{\omega\}) = \sum_{\omega \in A} p_\omega.$$

Therefore we have (a).

For (b), note that if $P(\{\omega\}) = p_\omega$, then by definition $p_\omega \geq 0$, and also

$$1 = P(\Omega) = P\left(\cup_{\omega \in \Omega}\{\omega\}\right) = \sum_{\omega \in \Omega} P(\{\omega\}) = \sum_{\omega \in \Omega} p_\omega.$$

For the converse, if the p_ω satisfy $p_\omega \geq 0$ and $\sum_{\omega \in \Omega} p_\omega = 1$, then we define a probability P by $P(A) \equiv \sum_{\omega \in A} p_\omega$, with the convention that an "empty" sum equals 0. Then $P(\emptyset) = 0$ and $P(\Omega) = \sum_{\omega \in \Omega} p_\omega = 1$. For countable additivity, it is trivial when Ω is finite; when Ω is countable it follows from the fact that one has the following associativity: $\sum_{i \in I} \sum_{\omega \in A_i} p_\omega = \sum_{\omega \in \cup_{i \in I} A_i} p_\omega$ if the A_i are pairwise disjoint. $\qquad \square$

Suppose first that Ω is finite. Any family of nonnegative terms summing up to 1 gives an example of a probability on Ω. But among all these examples the following is particularly important:

Definition 4.1. *A probability P on the finite set Ω is called* uniform *if $p_\omega = P(\{\omega\})$ does not depend on ω.*

In this case, it is immediate that

$$P(A) = \frac{\#(A)}{\#(\Omega)}.$$

Then computing the probability of any event A amounts to counting the number of points in A. On a given finite set Ω there is one and only one uniform probability.

We now give two examples which are important for applications.

a) The Hypergeometric distribution. An urn contains N white balls and M black balls. One draws n balls without replacement, so $n \leq N + M$. One gets X white balls and $n - X$ black balls. One is looking for the probability that $X = x$, where x is an arbitrary fixed integer.

Since we draw the balls without replacement, we can as well suppose that the n balls are drawn at once. So it becomes natural to consider that an outcome is a subset with n elements of the set $\{1, 2, \ldots, N+M\}$ of all $N+M$ balls (which can be assumed to numbered from 1 to $N+M$). That is, Ω is the family of all subsets with n points, and the total number of possible outcomes is $\#(\Omega) = \binom{N+M}{n} = \frac{(N+M)!}{n!(N+M-n)!}$: recall that for p and q two integers with $p \leq q$, then

$$p\,! = 1.2\ldots(p-1).p, \qquad \binom{q}{p} = \frac{q!}{p!(q-p)!}.$$

The quantity $\binom{q}{p}$, often pronounced "q choose p," can be thought of as the number of different ways to choose p items from q items, without regard to the order in which one chooses them.

Next, it is also natural to consider that all possible outcomes are equally likely, that is P is the uniform probability on Ω. The quantity X is a "random variable" because when the outcome ω is known, one also knows the number $X(\omega)$ of white balls which have been drawn. The set $X^{-1}(\{x\})$, also denoted by $\{X = x\}$, contains $\binom{N}{x}\binom{M}{n-x}$ points if $x \leq N$ and $n - x \leq M$, and it is empty otherwise. Hence

$$P(X = x) = \begin{cases} \dfrac{\binom{N}{x}\binom{M}{n-x}}{\binom{N+M}{n}} & \text{if } 0 \leq x \leq N \text{ and } 0 \leq n - x \leq M \\ 0 & \text{otherwise.} \end{cases}$$

We thus obtain, when x varies, the *distribution*, or the *law*, of X. This distribution is called the hypergeometric distribution, and it arises naturally in opinion polls: we have $N + M$ voters, among them N think "white" and M think "black", and one does a poll by asking the opinion of n voters (see Exercise 4.3 for an extension to more than 2 opinions).

b) The Binomial distribution. From the same urn as above, we draw n balls, but each time a ball is drawn we put it back, hence n can be as big as one wishes. We again want the probability $P(X = x)$, where x is an integer between 0 and n.

Here the natural probability space is the Cartesian product $\Omega = \Pi_{i=1}^{n} \Xi^i$ where $\Xi^i = \{1, 2, \ldots, N + M\}$ for all $i, 1 \leq i \leq n$; that is, $\Omega = \{1, 2, \ldots, N + M\}^n$, with again the uniform probability. Thus the number of elements in Ω, also called the **cardinality** of Ω and denoted $\#(\Omega)$, is given by $\#(\Omega) = (N + M)^n$. A simple computation shows that the cardinality of the set of all ω such that $X(\omega) = x$, that is $\#(X = x)$, equals $\binom{n}{x} N^x M^{n-x}$. Hence

$$P(X = x) = \binom{n}{x} \left(\frac{N}{N + M}\right)^x \left(\frac{M}{N + M}\right)^{n-x} \quad \text{for} \quad x = 0, 1, \ldots n.$$

Upon setting $p = \frac{N}{N+M}$, we usually write the result as follows:

$$P(X = x) = \binom{n}{x} p^x (1 - p)^{n-x} \quad \text{for} \quad x = 0, 1, \ldots n.$$

This formula gives the *Binomial distribution with size n and parameter p*. A priori p is arbitrary in $[0, 1]$ (in the previous example p is a rational number, but in general p may be any real number between 0 and 1). This distribution, which is ubiquitous in applications, is often denoted by $B(p, n)$.

c) The Binomial distribution as a limit of Hypergeometric distributions. In the situation a) above we now assume that n is fixed, while N and M increase to $+\infty$, in such a way that $\frac{N}{N+M}$ tends to a limit p (necessarily in $[0, 1]$). One easily checks that

$$P(X = x) \rightarrow \binom{n}{x} p^x (1 - p)^{n-x} \quad \text{for} \quad x = 0, 1, \ldots, n.$$

That is, the hypergeometric distributions "converge" to the binomial distribution $B(p, n)$. (Comparing with (b) above, the result is also intuitively evident, since when $N + M$ is big there is not much difference in drawing n balls with or without replacement).

Some examples with a countable state space.

1. The *Poisson distribution* of parameter $\lambda > 0$ is the probability P defined on $\mathbf{N}$ by
$$p_n = e^{-\lambda} \frac{\lambda^n}{n!}, \quad n = 0, 1, 2, 3, \ldots .$$

2. The *Geometric distribution* of parameter $\alpha \in [0, 1)$ is the probability defined on $\mathbf{N}$ by
$$p_n = (1 - \alpha)\alpha^n, \quad n = 0, 1, 2, 3, \ldots .$$

Note that in the Binomial model if n is large, then while in theory $\binom{n}{j} p^j (1-p)^{n-j}$ is known exactly, in practice it can be hard to compute. (Often it is beyond the capacities of quite powerful hand calculators, for example.) If n is large and p is small, however (as is often the case), there is an alternative method which we now describe.

Suppose p changes with n; call it p_n. Suppose further $\lim_{n \to \infty} np_n = \lambda$. One can show (see Exercise 4.1) that

$$\lim_{n \to \infty} \binom{n}{j} (p_n)^j (1-p_n)^{n-j} = e^{-\lambda} \frac{\lambda^j}{j!},$$

and thus one can easily approximate a Binomial probability (in this case) with a Poisson.

Appendix: Some useful result on series

In this Appendix we give a summary, mostly without proofs, of some useful result on series and summable families: these are primarily useful for studying probabilities on countable state spaces. These results (with proofs) can be found in most texts on Calculus (for example, see Chapter 10 of [18]).

First we establish some conventions. Quite often one is led to perform calculations involving $+\infty$ (written more simply as ∞) or $-\infty$. For these calculations to make sense we *always* use the following conventions:

$$+\infty + \infty = +\infty, \quad -\infty - \infty = -\infty, \quad a + \infty = +\infty, \quad a - \infty = -\infty \text{ if } a \in \mathbf{R},$$

$$0 \times \infty = 0, \quad a \in]0, \infty] \Rightarrow a \times \infty = +\infty, \quad a \in [-\infty, 0[\Rightarrow a \times \infty = -\infty.$$

Let u_n be a sequence of numbers, and consider the "partial sums" $S_n = u_1 + \ldots + u_n$.

S1: The series $\sum_n u_n$ is called *convergent* if S_n converges to a *finite* limit S, also denoted by $S = \sum_n u_n$ (the "sum" of the series).

S2: The series $\sum_n u_n$ is called *absolutely convergent* if the series $\sum_n |u_n|$ converges.

S3: If $u_n \geq 0$ for all n, the sequence S_n is increasing, hence always converges to a limit $S \in [0, \infty]$. We still write $S = \sum_n u_n$, although the series converges in the sense of (S1) if and only if $S < \infty$. The summands u_n can even take their values in $[0, \infty]$ provided we use the conventions above concerning addition with ∞.

In general the convergence of a series depends on the order in which the terms are enumerated. There are however two important cases where the ordering of the terms has no influence, and one speaks rather of "summable families" instead of "series" in these cases, which are S4 and S5 below:

S4:   When the u_n are reals and the series is absolutely convergent one can modify the order in which the terms are taken without changing the absolute convergence, nor the sum of the series.

S5:   When $u_n \in [0, \infty]$ for all n, the sum $\sum_n u_n$ (which is finite or infinite: cf. (S3) above) does not change if the order is changed.

S6:   When $u_n \in [0, \infty]$, or when the series is absolutely convergent, we have the following associativity property: let $(A_i)_{i \in I}$ be a partition of $\mathbf{N}^*$, with $I = \{1, 2, \ldots, N\}$ for some integer N, or $I = \mathbf{N}^*$. For each $i \in I$ we set $v_i = \sum_{n \in A_i} u_n$: if A_i is finite this is an ordinary sum, otherwise v_i is itself the sum of a series. Then we have $\sum_n u_n = \sum_{i \in I} v_i$ (this latter sum is again the sum of a series if $I = \mathbf{N}^*$).

Exercises for Chapter 4

4.1 (Poisson Approximation to the Binomial) Let P be a Binomial probability with probability of success p and number of trials n. Let $\lambda = pn$. Show that

$$P(k \text{ successes})$$
$$= \frac{\lambda^k}{k!} \left(1 - \frac{\lambda}{n}\right)^n \left\{ \left(\frac{n}{n}\right) \left(\frac{n-1}{n}\right) \cdots \left(\frac{n-k+1}{n}\right) \right\} \left(1 - \frac{\lambda}{n}\right)^{-k}.$$

Let $n \to \infty$ and let p change so that λ remains constant. Conclude that for small p and large n,

$$P(k \text{ successes}) \approx \frac{\lambda^k}{k!} e^{-\lambda}, \quad \text{where } \lambda = pn.$$

[Note: In general for this approximation technique to be good one needs n large, p small, and also $\lambda = np$ to be of moderate size — for example $\lambda \leq 20$.]

4.2 (Poisson Approximation to the Binomial, continued) In the setting of Exercise 4.1, let $p_k = P(\{k\})$ and $q_k = 1 - p_k$. Show that the q_k are the probabilities of singletons for a Binomial distribution $B(1 - p, n)$. Deduce a Poisson approximation of the Binomial when n is large and p is close to 1.

4.3 We consider the setting of the hypergeometric distribution, except that we have m colors and N_i balls of color i. Set $N = N_1 + \ldots + N_m$, and call X_i the number of balls of color i drawn among n balls. Of course $X_1 + \ldots + X_m = n$. Show that

$$P(X_1 = x_1, \ldots, X_m = x_m) = \begin{cases} \dfrac{\binom{N_1}{x_1} \cdots \binom{N_m}{x_m}}{\binom{N}{n}} & \text{if } x_1 + \ldots + x_m = n \\ 0 & \text{otherwise.} \end{cases}$$

5 Random Variables on a Countable Space

In Chapter 5 we again assume Ω is countable and $\mathcal{A} = 2^\Omega$. A *random variable* X in this case is defined to be a function from Ω into a set T. A random variable represents an unknown quantity (hence the term variable) that varies not as a variable in an algebraic relation (such as $x^2 - 9 = 0$), but rather varies with the outcome of a random event. Before the random event, we know which values X could possibly assume, but we do not know which one it will take until the random event happens. This is analogous to algebra when we know that x can take on *a priori* any real value, but we do not know which one (or ones) it will take on until we solve the equation $x^2 - 9 = 0$ (for example).

Note that even if the state space (or range space) T is not countable, the image T' of Ω under X (that is, all points $\{i\}$ in T for which there exists an $\omega \in \Omega$ such that $X(\omega) = i$) is either finite or countably infinite.

We can then define the *distribution of X* (also called the *law of X*) on the range space T' of X by

$$P^X(A) = P(\{\omega : X(\omega) \in A\}) = P(X^{-1}(A)) = P(X \in A).$$

That this formula defines a Probability measure on T' (with the σ-algebra $2^{T'}$ of all subsets of T') is evident. Since T' is at most countable, this probability is completely determined by the following numbers:

$$p_j^X = P(X = j) = \sum_{\{\omega : X(\omega) = j\}} p_\omega.$$

Sometimes, the family $(p_j^X : j \in T')$ is also called the distribution (or the law) of X. We have of course $P_X(A) = \sum_{j \in A} p_j^X$. If P^X has a known distribution, for example Poisson, then we say that X is a Poisson random variable.

Definition 5.1. *Let X be a real-valued random variable on a countable space Ω. The expectation of X, denoted $E\{X\}$, is defined to be*

$$E\{X\} = \sum_\omega X(\omega) p_\omega,$$

provided this sum makes sense: this is the case when Ω is finite; this is also the case when Ω is countable, when the series is absolutely convergent or

$X \geq 0$ *always* (in the latter case, the above sum and hence $E\{X\}$ as well may take the value $+\infty$).

This definition can be motivated as follows: If one repeats an experiment n times, and one records the values $X_1, X_2, \ldots, X_n$ of X corresponding to the n outcomes, then the *empirical mean* $\frac{1}{n}(X_1 + \ldots + X_n)$ is $\sum_{\omega \in \Omega} X(\omega) f_n(\{\omega\})$, where $f_n(\{\omega\})$ denotes the frequency of appearance of the singleton $\{\omega\}$. Since $f_n(\{\omega\})$ "converges" to $P(\{\omega\})$, it follows (at least when Ω is finite) that the empirical mean converges to the expectation $E\{X\}$ as defined above.

Define $\mathcal{L}^1$ to be the space of real valued random variables on $(\Omega, \mathcal{A}, P)$ which have a finite expectation.

The following facts follow easily:

(i) $\mathcal{L}^1$ is a vector space, and the expectation operator E is linear,
(ii) the expectation operator E is positive: if $X \in \mathcal{L}^1$ and $X \geq 0$, then $E\{X\} \geq 0$. More generally if $X, Y \in \mathcal{L}^1$ and $X \leq Y$ then $E\{X\} \leq E\{Y\}$.
(iii) $\mathcal{L}^1$ contains all bounded random variables. If $X \equiv a$, then $E\{X\} = a$.
(iv) If $X \in \mathcal{L}^1$, its expectation depends only on its distribution and, if T' is the range of X,

$$E\{X\} = \sum_{j \in T'} j P(X = j). \tag{5.1}$$

(v) If $X = 1_A$ is the indicator function of an event A, then $E\{X\} = P(A)$.

We observe that if $\sum_\omega (X(\omega))^2 p_\omega$ is absolutely convergent, then

$$\sum_\omega |X(\omega)| p_\omega \leq \sum_{|X(\omega)| < 1} X(\omega) p_\omega + \sum_{|X(\omega)| \geq 1} X(\omega) p_\omega$$
$$\leq \sum_\omega p_\omega + \sum_\omega (X(\omega))^2 p_\omega < \infty,$$

and $X \in \mathcal{L}^1$ too.

An important family of inequalities involving expectation follow from the next theorem.

Theorem 5.1. *Let $h : \mathbf{R} \to [0, \infty)$ be a nonnegative function and let X be a real valued random variable. Then*

$$P\{\omega : h(X(\omega)) \geq a\} \leq \frac{E\{h(X)\}}{a}$$

for all $a > 0$.

Proof. Since X is an r.v. so also is $Y = h(X)$; let

$$A = Y^{-1}([a, \infty)) = \{\omega : h(X(\omega)) \geq a\} = \{h(X) \geq a\}.$$

Then $h(X) \geq a 1_A$, hence

$$E\{h(X)\} \geq E\{a 1_A\} = a E\{1_A\} = a P(A)$$

and we have the result. $\square$

Corollary 5.1 (Markov's Inequality).

$$P\{|X| \geq a\} \leq \frac{E\{|X|\}}{a}.$$

Proof. Take $h(x) = |x|$ in Theorem 5.1. □

Definition 5.2. *Let X be a real-valued random variable with X^2 in $\mathcal{L}^1$. The* Variance *of X is defined to be*

$$\sigma^2 = \sigma_X^2 \equiv E\{(X - E(X))^2\}.$$

The standard deviation *of X, σ_X, is the nonnegative square root of the variance.* The primary use of the standard deviation is to report statistics in the correct (and meaningful) units.

An example of the problem units can pose is as follows: let X denote the number of children in a randomly chosen family. Then the units of the variance will be "square children", whereas the units for the standard deviation σ_X will be simply "children".

If $E\{X\}$ represents the expected, or average, value of X (often called the *mean*), then $E\{|X - E(X)|\} = E\{|X - \mu|\}$ where $\mu = E\{X\}$, represents the average difference from the mean, and is a measure of how "spread out" the values of X are. Indeed, it measures how the values *vary* from the mean. The variance is the average squared distance from the mean. This has the effect of diminishing small deviations from the mean and enlarging big ones. However the variance is usually easier to compute than is $E\{|X - \mu|\}$, and often it has a simpler expression. (See for example Exercise 5.11.) The variance too can be thought of as a measure of variability of the random variable X.

Corollary 5.2 (Chebyshev's Inequality). *If X^2 is in $\mathcal{L}^1$, then we have*

$$\text{(a) } P\{|X| \geq a\} \leq \frac{E\{X^2\}}{a^2} \qquad \text{for} \quad a > 0,$$

$$\text{(b) } P\{|X - E\{X\}| \geq a\} \leq \frac{\sigma_X^2}{a^2} \qquad \text{for} \quad a > 0.$$

Proof. Both inequalities are known as Chebyshev's inequality. For part (a), take $h(x) = x^2$ and then by Theorem 5.1

$$P\{|X| \geq a\} = P\{h(X) \geq a^2\} \leq \frac{E\{X^2\}}{a^2}.$$

For part (b), let $Y = |X - E\{X\}|$. Then

$$P\{|X - E\{X\}| \geq a\} = P\{Y \geq a\} = P\{Y^2 \geq a^2\} \leq \frac{E\{Y^2\}}{a^2} = \frac{\sigma_X^2}{a^2}.$$

□

Corollary 5.2 is also known as the Bienaymé-Chebyshev inequality.

Examples:

1) X is *Poisson* with parameter λ. Then $X: \Omega \to \mathbf{N}$ (the natural numbers), and

$$P(X \in A) = \sum_{j \in A} P(X = j) = \sum_{j \in A} \frac{\lambda^j}{j!} e^{-\lambda}.$$

The expectation of X is

$$E\{X\} = \sum_{j=0}^{\infty} j P(X = j) = \sum_{j=0}^{\infty} j \frac{\lambda^j}{j!} e^{-\lambda}$$

$$= \lambda \sum_{j=1}^{\infty} \frac{\lambda^{j-1}}{(j-1)!} e^{-\lambda} = \lambda e^{\lambda} e^{-\lambda} = \lambda.$$

2) X has the *Bernoulli distribution* if X takes on only two values: 0 and 1. X corresponds to an experiment with only two outcomes, usually called "success" and "failure". Usually $\{X = 1\}$ corresponds to "success". Also it is customary to call $P(\{X = 1\}) = p$ and $P(\{X = 0\}) = q = 1 - p$. Note

$$E\{X\} = 1P(X = 1) + 0P(X = 0) = 1.p + 0.q = p.$$

3) X has the *Binomial distribution* if P^X is the Binomial probability. That is, for a given and fixed n, X can take on the values $\{0, 1, 2, \ldots, n\}$.

$$P(\{X = k\}) = \binom{n}{k} p^k (1 - p)^{n-k},$$

where $0 \leq p \leq 1$ is fixed.
Suppose we perform a success/failure experiment n times independently. Let

$$Y_i = \begin{cases} 1 \text{ if success on the } i^{\text{th}} \text{ trial,} \\ 0 \text{ if failure on the } i^{\text{th}} \text{ trial.} \end{cases}$$

Then $X = Y_1 + \ldots + Y_n$ has the Binomial distribution (see Chapter 4). That is, a Binomial random variable is the sum of n Bernoulli random variables. Therefore

$$E\{X\} = E\left\{\sum_{i=1}^{n} Y_i\right\} = \sum_{i=1}^{n} E\{Y_i\} = \sum_{i=1}^{n} p = np.$$

Note that we could also have computed $E\{X\}$ combinatorially by using the definition:

$$E\{X\} = \sum_{i=0}^{n} i P(X = i) = \sum_{i=1}^{\infty} i \binom{n}{i} p^i (1 - p)^{n-i},$$

but this would have been an unpleasant calculation.

4) Suppose we are performing repeated independent Bernoulli trials. If instead of having a fixed number n of trials to be chosen in advance, suppose we keep performing trials until we have achieved a given number of success. Let X denote the number of failures before we reach a success. X has a *Geometric distribution*, with parameter $1 - p$:

$$P(X = k) = (1 - p)^k p, \qquad k = 0, 1, 2, 3, \ldots$$

where p is the probability of success. We then have

$$E\{X\} = \sum_{k=0}^{\infty} k P(X = k) = \sum_{k=0}^{\infty} k p(1 - p)^k = p(1 - p)\frac{1}{p^2} = \frac{1 - p}{p}.$$

5) In the same framework as (4), if we continue independent Bernoulli trials until we achieve the rth success, then we have *Pascal's distribution*, also known as the *Negative Binomial distribution*. We say X has the Negative Binomial distribution with parameters r and p if

$$P(X = j) = \binom{j + r - 1}{r - 1} p^r (1 - p)^j$$

for $j = 0, 1, 2, \ldots$. X represents the number of failures that must be observed before r successes are observed. If one is interested in the *total* number of trials required, call that r.v. Y, then $Y = X + r$.
Note that if X is Negative Binomial, then

$$X = \sum_{i=1}^{r} Z_i,$$

where Z_i are geometric random variables with parameter $1 - p$. Therefore

$$E\{X\} = \sum_{i=1}^{r} E\{Z_i\} = \frac{r(1 - p)}{p}.$$

6) A distribution common in the social sciences is the *Pareto distribution*, also known as the *Zeta distribution*. Here X takes its values in $\mathbf{N}^*$, where

$$P(X = j) = c\frac{1}{j^{\alpha+1}}, \qquad j = 1, 2, 3, \ldots$$

for a *fixed parameter* $\alpha > 0$. The constant c is such that $c \sum_{j=1}^{\infty} \frac{1}{j^{\alpha+1}} = 1$. The function

$$\zeta(s) = \sum_{k=1}^{\infty} \frac{1}{k^s}, \qquad s > 1,$$

is known as the *Riemann zeta function*, and it is extensively tabulated. Thus $c = \frac{1}{\zeta(\alpha+1)}$, and

$$P(X = j) = \frac{1}{\zeta(\alpha + 1)} \frac{1}{j^{\alpha+1}}.$$

The mean is easily calculated in terms of the Riemann zeta function:

$$E\{X\} = \sum_{j=1}^{\infty} jP(X = j) = \frac{1}{\zeta(\alpha + 1)} \sum_{j=1}^{\infty} \frac{j}{j^{\alpha+1}}$$

$$= \frac{1}{\zeta(\alpha + 1)} \sum_{j=1}^{\infty} \frac{1}{j^{\alpha}} = \frac{\zeta(\alpha)}{\zeta(\alpha + 1)}.$$

7) If the state space E of a random variable X has only a finite number of points, say n, and each point is equally likely, then X is said to have a uniform distribution. In the case where

$$P(X = j) = \frac{1}{n}, \qquad j = 1, 2, \ldots, n,$$

then X has the *Discrete Uniform* distribution with parameter n. Using that $\sum_{i=1}^{n} i = \frac{n(n+1)}{2}$, we have

$$E\{X\} = \sum_{j=1}^{n} jP(X = j) = \sum_{j=1}^{n} j\frac{1}{n} = \frac{1}{n} \sum_{j=1}^{n} j = \frac{n(n+1)}{n \cdot 2} = \frac{n+1}{2}.$$

Exercises for Chapter 5

5.1 Let $g : [0, \infty) \to [0, \infty)$ be strictly increasing and nonnegative. Show that

$$P(|X| \geq a) \leq \frac{E\{g(|X|)\}}{g(a)} \qquad \text{for } a > 0.$$

5.2 Let $h : \mathbf{R} \to [0, \alpha]$ be a nonnegative (bounded) function. Show that for $0 \leq a < \alpha$,

$$P\{h(X) \geq a\} \geq \frac{E\{h(X)\} - a}{\alpha - a}.$$

5.3 Show that $\sigma_X^2 = E\{X^2\} - E\{X\}^2$, assuming both expectations exist.

5.4 Show that $E\{X\}^2 \leq E\{X^2\}$ always, assuming both expectations exist.

5.5 Show that $\sigma_X^2 = E\{X(X-1)\} + \mu_X - \mu_X^2$, where $\mu_X = E\{X\}$, assuming all expectations exist.

5.6 Let X be Binomial $B(p, n)$. For what value of j is $P(X = j)$ the greatest? (*Hint:* Calculate $\frac{P(X=k)}{P(X=k-1)}$.)
[Ans.: $[(n + 1)p]$, where $[x]$ denotes integer part of x.]

5.7 Let X be Binomial $B(p, n)$. Find the probability X is even. [Ans.: $\frac{1}{2}(1 + (1 - 2p)^n)$.]

5.8 Let X_n be Binomial $B(p_n, n)$ with $\lambda = np_n$ being constant. Let $A_n = \{X_n \geq 1\}$, and let Y be Poisson (λ). Show that $\lim_{n \to \infty} P(X_n = j \mid A_n) = P(Y = j \mid Y \geq 1)$.

5.9 Let X be Poisson (λ). What value of j maximizes $P(X = j)$?
[Ans.: $[\lambda]$.] (*Hint:* See Exercise 5.6.)

5.10 Let X be Poisson (λ). For fixed $j > 0$, what value of λ maximizes $P(X = j)$?
[Ans.: j.]

5.11 Let X be Poisson (λ) with λ a positive integer. Show $E\{|X - \lambda|\} = \frac{2\lambda^\lambda e^{-\lambda}}{(\lambda - 1)!}$, and that $\sigma_X^2 = \lambda$.

5.12 * Let X be Binomial $B(p, n)$. Show that for $\lambda > 0$ and $\varepsilon > 0$,

$$P(X - np > n\varepsilon) \leq E\{\exp(\lambda(X - np - n\varepsilon))\}.$$

5.13 Let X_n be Binomial $B(p, n)$ with $p > 0$ fixed. Show that for any fixed $b > 0$, $P(X_n \leq b)$ tends to 0.

5.14 Let X be Binomial $B(p, n)$ with $p > 0$ fixed, and $a > 0$. Show that

$$P\left(\left|\frac{X}{n} - p\right| > a\right) \leq \frac{\sqrt{p(1-p)}}{a^2 n} \min\left\{\sqrt{p(1-p)}, a\sqrt{n}\right\},$$

and also that $P(|X - np| \leq n\varepsilon)$ tends to 1 for all $\varepsilon > 0$.

5.15 * Let X be a Binomial $B(\frac{1}{2}, n)$, where $n = 2m$. Let

$$a(m, k) = \frac{4^m}{\binom{2m}{m}} P(X = m + k).$$

Show that $\lim_{m \to \infty} (a(m, k))^m = e^{-k^2}$.

5.16 Let X be Geometric. Show that for $i, j > 0$,

$$P(X > i + j \mid X > i) = P(X > j).$$

5.17 Let X be Geometric (p). Show

$$E\left\{\frac{1}{1+X}\right\} = \log((1-p)^{\frac{p}{p-1}}).$$

5.18 A coin is tossed independently and repeatedly with the probability of heads equal to p.

a) What is the probability of only heads in the first n tosses?
b) What is the probability of obtaining the first tail at the nth toss?
c) What is the expected number of tosses required to obtain the first tail?
   [Ans.: $\frac{1}{1-p}$.]

5.19 Show that for a sequence of events $(A_n)_{n \geq 1}$,

$$E\left\{\sum_{n=1}^{\infty} 1_{A_n}\right\} = \sum_{n=1}^{\infty} P(A_n),$$

where ∞ is a possible value for each side of the equation.

5.20 Suppose X takes all its values in $\mathbf{N} (= \{0, 1, 2, 3, \ldots\})$. Show that

$$E\{X\} = \sum_{n=0}^{\infty} P(X > n).$$

5.21 Let X be Poisson (λ). Show for $r = 2, 3, 4, \ldots,$

$$E\{X(X-1)\ldots(X-r+1)\} = \lambda^r.$$

5.22 Let X be Geometric (p). Show for $r = 2, 3, 4, \ldots,$

$$E\{X(X-1)\ldots(X-r+1)\} = \frac{r!p^r}{(1-p)^r}.$$

6 Construction of a Probability Measure

Here we no longer assume Ω is countable. We assume given Ω and a σ-algebra $\mathcal{A} \subset 2^\Omega$. $(\Omega, \mathcal{A})$ is called a *measurable space*. We want to construct probability measures on $\mathcal{A}$. When Ω is finite or countable we have already seen this is simple to do. When Ω is uncountable, the same technique does not work; indeed, a "typical" probability P will have $P(\{\omega\}) = 0$ for all ω, and thus the family of all numbers $P(\{\omega\})$ for $\omega \in \Omega$ does *not* characterize the probability P in general.

It turns out in many "concrete" situations – in particular in the next chapter – that it is often relatively simple to construct a "probability" on an algebra which generates the σ-algebra $\mathcal{A}$, and the problem at hand is then to extend this probability to the σ-algebra itself. So, let us suppose $\mathcal{A}$ is the σ-algebra generated by an algebra $\mathcal{A}_0$, and let us further suppose we are given a probability P on the algebra $\mathcal{A}_0$: that is, a function $P : \mathcal{A}_0 \to [0, 1]$ satisfying

1. $P(\Omega) = 1$.
2. (Countable Additivity) for any sequence (A_n) of elements of $\mathcal{A}_0$, pairwise disjoint, and such that $\cup_n A_n \in \mathcal{A}_0$, we have $P(\cup_n A_n) = \sum_n P(A_n)$.

It might seem natural to use for $\mathcal{A}$ the set of all subsets of Ω, as we did in the case where Ω was countable. We do not do so for the following reason, illustrated by an example: suppose $\Omega = [0, 1]$, and let us define a set function P on intervals of the form $P((a, b]) = b - a$, where $0 \leq a \leq b \leq 1$. This is a natural "probability measure" that assigns the usual length of an interval as its probability. Suppose we want to extend P in a unique way to $2^{[0,1]} = $ all subsets of $[0, 1]$ such that (i) $P(\Omega) = 1$; and (ii) $P(\cup_{n=1}^\infty A_n) = \sum_{n=1}^\infty P(A_n)$ for any sequence of subsets $(A_n)_{n \geq 1}$ with $A_n \cap A_m = \phi$ for $n \neq m$; then one can *prove* that no such P exists! The collection of sets $2^{[0,1]}$ is simply too big for this to work. Borel realized that we can however do this on a smaller collection of sets, namely the smallest σ-algebra containing intervals of the form $(a, b]$. This is the import of the next theorem:

Theorem 6.1. *Each probability P defined on the algebra $\mathcal{A}_0$ has a* unique *extension (also called P) on $\mathcal{A}$.*

We will show only the uniqueness. For the existence on can consult any standard text on measure theory; for example [16] or [23]. First we need to establish a very useful theorem.

Definition 6.1. *A class C of subsets of Ω is* closed under finite intersections *if for when $A_1, \ldots, A_n \in C$, then $A_1 \cap A_2 \cap \ldots \cap A_n \in C$ as well (n arbitrary but finite).*

A class C is closed under increasing limits *if wherever $A_1 \subset A_2 \subset A_3 \subset \ldots \subset A_n \subset \ldots$ is a sequence of events in C, then $\cup_{n=1}^{\infty} A_n \in C$ as well.*

A class C is closed under differences *if whenever $A, B \in C$ with $A \subset B$, then $B \setminus A \in C$[1].*

Theorem 6.2 (Monotone Class Theorem). *Let C be a class of subsets of Ω, closed under finite intersections and containing Ω. Let B be the smallest class containing C which is closed under increasing limits and by difference. Then $B = \sigma(C)$.*

Proof. First note that the intersection of classes of sets closed under increasing limits and differences is again a class of that type. So, by taking the intersection of all such classes, there always exists a *smallest* class containing C which is closed under increasing limits and by differences. For each set B, denote B_B to be the collection of sets A such that $A \in B$ and $A \cap B \in B$. Given the properties of B, one easily checks that B_B is closed under increasing limits and by difference.

Let $B \in C$; for each $C \in C$ one has $B \cap C \in C \subset B$ and $C \in B$, thus $C \in B_B$. Hence $C \subset B_B \subset B$. Therefore $B = B_B$, by the properties of B and of B_B.

Now let $B \in B$. For each $C \in C$, we have $B \in B_C$, and because of the preceding, $B \cap C \in B$, hence $C \in B_B$, whence $C \subset B_B \subset B$, hence $B = B_B$.

Since $B = B_B$ for all $B \in B$, we conclude B is closed by finite intersections. Furthermore $\Omega \in B$, and B is closed by difference, hence also under complementation. Since B is closed by increasing limits as well, we conclude B is a σ-algebra, and it is clearly the smallest such containing C.   □

The proof of the uniqueness in Theorem 6.1 is an immediate consequence of Corollary 6.1 below, itself a consequence of the Monotone Class Theorem.

Corollary 6.1. *Let P and Q be two probabilities defined on A, and suppose P and Q agree on a class $C \subset A$ which is closed under finite intersections. If $\sigma(C) = A$, we have $P = Q$.*

Proof. $\Omega \in A$ because A is a σ-algebra, and since $P(\Omega) = Q(\Omega) = 1$ because they are both probabilities, we can assume without loss of generality that $\Omega \subset C$. Let $B = \{A \in A : P(A) = Q(A)\}$. By the definition of a Probability measure and Theorem 2.3, B is closed by difference and by increasing limits.

---

[1] $B \setminus A$ denotes $B \cap A^c$

Also $\mathcal{B}$ contains $\mathcal{C}$ by hypothesis. Therefore since $\sigma(\mathcal{C}) = \mathcal{A}$, we have $\mathcal{B} = \mathcal{A}$ by the Monotone Class Theorem (Theorem 6.2). $\qquad\square$

There is a version of Theorem 6.2 for functions. We will not have need of it in this book, but it is a useful theorem to know in general so we state it here without proof. For a proof the reader can consult [19, p. 365]. Let $\mathcal{M}$ be a class of functions mapping a given space Ω into $\mathbf{R}$. We let $\sigma(\mathcal{M})$ denote the smallest σ-algebra on Ω that makes all of the functions in M measurable:

$$\sigma(\mathcal{M}) = \{f^{-1}(\wedge); \wedge \in \mathcal{B}(\mathbf{R}); f \in M\}.$$

Theorem 6.3 (Monotone Class Theorem). *Let $\mathcal{M}$ be a class of bounded functions mapping Ω into $\mathbf{R}$. Suppose $\mathcal{M}$ is closed under multiplication: $f, g \in \mathcal{M}$ implies $fg \in \mathcal{M}$. Let $\mathcal{A} = \sigma(\mathcal{M})$. Let $\mathcal{H}$ be a vector space of functions with $\mathcal{H}$ containing $\mathcal{M}$. Suppose $\mathcal{H}$ contains the constant functions and is such that whenever $(f_n)_{n \geq 1}$ is a sequence in $\mathcal{H}$ such that $0 \leq f_1 \leq f_2 \leq f_3 \leq \ldots$, then if $f = \lim_{n \to \infty} f_n$ is bounded, then f is in $\mathcal{H}$. Then $\mathcal{H}$ contains all bounded, $\mathcal{A}$-measurable functions.*

It is possible to do a little bit better that in Theorem 6.1: we can extend the probability to a σ-algebra slightly bigger than $\mathcal{A}$. This is the aim of the rest of this chapter.

Definition 6.2. *Let P be a probability on $\mathcal{A}$. A* null set *(or negligible set) for P is a subset A of Ω such that there exists a $B \in \mathcal{A}$ satisfying $A \subset B$ and $P(B) = 0$.*

We say that a property holds *almost surely* (a.s. in short) if it holds outside a negligible set. This notion clearly depends on the probability, so we say sometimes P-almost surely, or P-a.s.

The negligible sets are not necessarily in $\mathcal{A}$. Nevertheless it is natural to say that they have probability zero. In fact one even can extend the probability to the σ-algebra which is generated by $\mathcal{A}$ and all P-negligible sets: this is what is done in the following theorem, which we do not explicitly use in the sequel.

Theorem 6.4. *Let P be a probability on $\mathcal{A}$ and let $\mathcal{N}$ be the class of all P-negligible sets. Then $\mathcal{A}' = \{A \cup N : A \in \mathcal{A}, N \in \mathcal{N}\}$ is a σ-algebra, called the P-completion of $\mathcal{A}$. This is the smallest σ-algebra containing $\mathcal{A}$ and $\mathcal{N}$, and P extends uniquely as a probability (still denoted by P) on $\mathcal{A}'$, by setting $P(A \cup N) = P(A)$ for $A \in \mathcal{A}$ and $N \in \mathcal{N}$.*

Proof. The uniqueness of the extension is straightforward, provided it exists. Also, since $\emptyset$ belongs to both $\mathcal{A}$ and $\mathcal{N}$, the fact that $\mathcal{A}'$ is the smallest σ-algebra containing $\mathcal{A}$ and $\mathcal{N}$ is trivial, provided $\mathcal{A}'$ is a σ-algebra. Hence it is enough to prove that $\mathcal{A}'$ is a σ-algebra and that, if we set $Q(B) = P(A)$ for

$B = A \cup N$ (with $A \in \mathcal{A}$ and $N \in \mathcal{N}$), then $Q(B)$ does not depend on the decomposition $B = A \cup N$ and Q is a probability on $\mathcal{A}'$.

First, $\Omega \in \mathcal{A} \subset \mathcal{A}'$ and we have seen already that $\emptyset \in \mathcal{A}'$. Next, since $\mathcal{A}$ and $\mathcal{N}$ are stable by countable unions (to verify this for $\mathcal{N}$, use subadditivity (cf. Exercise 2.15)), $\mathcal{A}'$ is also stable by countable unions. Let $B = A \cup N \in \mathcal{A}'$, with $A \in \mathcal{A}$ and $N \in \mathcal{N}$. There exists $C \in \mathcal{A}$ with $P(C) = 0$ and $N \subset C$. Set $A' = A^c \cap C^c$ (which belongs to $\mathcal{A}$) and $N' = N^c \cap A^c \cap C$ (which is contained in C and thus belongs to $\mathcal{N}$); since $B^c = A' \cup N'$, we have $B^c \in \mathcal{A}'$, and $\mathcal{A}'$ is stable by complementation: hence $\mathcal{A}'$ is a σ-algebra.

Suppose now that $A_1 \cup N_1 = A_2 \cup N_2$ with $A_i \in \mathcal{A}$ and $N_i \in \mathcal{N}$. The symmetrical difference $A_1 \triangle A_2 = (A_1 \cap A_2^c) \cup (A_1^c \cap A_2)$ is contained in $N_1 \cup N_2$, which itself is an element C of $\mathcal{A}$ with zero probability: hence $P(A_1) = P(A_2)$, which shows that Q is defined unambiguously, and obviously coincide with P on $\mathcal{A}$. Finally, the fact that Q is a probability on $\mathcal{A}'$ is evident. $\qquad \square$

7 Construction of a Probability Measure on R

This chapter is an important special case of what we dealt with in Chapter 6. We assume that $\Omega = \mathbf{R}$.

Let $\mathcal{B}$ be the Borel σ-algebra of $\mathbf{R}$.. (That is, $\mathcal{B} = \sigma(\mathcal{O})$, where $\mathcal{O}$ are the open subsets of $\mathbf{R}$.)

Definition 7.1. *The distribution function induced by a probability P on $(\mathbf{R}, \mathcal{B})$ is the function*

$$F(x) = P((-\infty, x]). \tag{7.1}$$

Theorem 7.1. *The distribution function F characterizes the probability.*

Proof. We want to show that knowledge of F defined by (7.1) uniquely determines P. That is, if there is another probability Q such that

$$G(x) = Q((-\infty, x])$$

for $x \in \mathbf{R}$, and if $F = G$, then also $P = Q$.

We begin by letting $\mathcal{B}_0$ be the set of finite disjoint unions of intervals of the form $(x, y]$, with $-\infty \leq x \leq y \leq +\infty$ (with the convention that $(x, \infty] = (x, \infty)$; observe also that $(x, y] = \emptyset$ if $x = y$). It is easy to see that $\mathcal{B}_0$ is an algebra. Moreover if (a, b) is an open interval, then $(a, b) = \cup_{n=N}^{\infty}(a, b - \frac{1}{n}]$, for some N large enough, so $\sigma(\mathcal{B}_0)$ contains all open intervals. But all open sets on the line can be expressed as countable unions of open intervals, and since the Borel sets $(= \mathcal{B})$ are generated by the open sets, $\sigma(\mathcal{B}_0) \supset \mathcal{B}$ (note also that $\cap_{n=1}^{\infty}(a, b + \frac{1}{n}) = (a, b]$, so $\mathcal{B}_0 \subset \mathcal{B}$ and thus $\mathcal{B} = \sigma(\mathcal{B}_0)$).

The relation (7.1) implies that

$$P((x, y]) = F(y) - F(x),$$

and if $A \in \mathcal{B}_0$ is of the form

$$A = \cup_{1 \leq i \leq n}(x_i, y_i] \text{ with } y_i < x_{i+1},$$

then $P(A) = \sum_{1 \leq i \leq n}\{F(y_i) - F(x_i)\}$.

If Q is another probability measure such that

$$F(x) = Q((-\infty, x]),$$

then the preceding shows that $P = Q$ on $\mathcal{B}_0$. Theorem 6.1 then implies that $P = Q$ on all of $\mathcal{B}$, so they are the same Probability measure. □

The significance of Theorem 7.1 is that we know, in principle, the complete probability measure P if we know its distribution function F : that is, we can in principle determine from F the probability $P(A)$ for any given Borel set A. (Determining these probabilities in practice is another matter.)

It is thus important to characterize *all* functions F which are distribution functions, and also to construct them easily. (Recall that a function F is *right continuous* if $\lim_{y \downarrow x} F(y) = F(x)$, for all $x \in \mathbf{R}$.)

Theorem 7.2. *A function F is the distribution function of a (unique) probability on $(\mathbf{R}, \mathcal{B})$ if and only if one has:*

(i) *F is non-decreasing;*
(ii) *F is right continuous;*
(iii) *$\lim_{x \to -\infty} F(x) = 0$ and $\lim_{x \to +\infty} F(x) = 1$.*

Proof. Assume that F is a distribution function. If $y > x$, then $(-\infty, x] \subset (-\infty, y]$, so $P((-\infty, x]) \leq P((-\infty, y])$ and thus $F(x) \leq F(y)$. Thus we have (i). Next let x_n decrease to x. Then $\cap_{n=1}^{\infty}(-\infty, x_n] = (-\infty, x]$, and the sequence of events $\{(-\infty, x_n]; n \geq 1\}$ is a decreasing sequence. Therefore $P(\cap_{n=1}^{\infty}(-\infty, x_n]) = \lim_{n \to \infty} P((-\infty, x_n]) = P((-\infty, x])$ by Theorem 2.3, and we have (ii). Similarly, Theorem 2.3 gives us (iii) as well.

Next we assume that we have (i), (ii), and (iii) and we wish to show F is a distribution function. In accordance with (iii), let us set $F(-\infty) = 0$ and $F(+\infty) = 1$. As in the proof of Theorem 7.1, let $\mathcal{B}_0$ be the set of finite disjoint unions of intervals of the form $(x, y]$, with $-\infty \leq x < y \leq +\infty$. Define a set function P, $P : \mathcal{B}_0 \to [0, 1]$ as follows: for $A = \cup_{1 \leq i \leq n}(x_i, y_i]$ with $y_i < x_{i+1}$,

$$P(A) \equiv \sum_{1 \leq i \leq n} \{F(y_i) - F(x_i)\}.$$

(Note that since $y_i < x_{i+1}$, there is a unique way to represent a set $A \in \mathcal{B}_0$, except for $A = \emptyset$, for which the above formula trivially gives $P(\emptyset) = 0$ whatever representation is used.) Condition (iii) gives us that $P(\mathbf{R}) = 1$. We wish to show that P is a probability on the algebra $\mathcal{B}_0$; if we can show that, we know by Theorem 6.1 that it has a unique extension to all of $\mathcal{B}$, and Theorem 7.2 will be proved.

To show P is a probability on $\mathcal{B}_0$ we must establish countable additivity: that is, for any sequence $(A_n) \in \mathcal{B}_0$, pairwise disjoint, and such that $\cup_{n=1}^{\infty} A_n \in \mathcal{B}_0$, then $P(\cup_{n=1}^{\infty} A_n) = \sum_{n=1}^{\infty} P(A_n)$. Now, finite additivity (that is, for any finite family $A_1, \ldots, A_n \in \mathcal{B}_0$, pairwise disjoint, then $P(\cup_{i=1}^{n} A_i) = \sum_{i=1}^{n} P(A_i)$) is a trivial consequence of the definition of P, and Theorem 2.3 gives equivalent conditions for a finitely additive P to be a probability on a σ-algebra $\mathcal{B}$; however it is simple to check that these conditions are also equivalent on an algebra $\mathcal{B}_0$. Thus it suffices to show that if $A_n \in \mathcal{B}_0$ with A_n decreasing to $\emptyset$ (the empty set), then $P(A_n)$ decreases to 0.

Let $A_n \in \mathcal{B}_0$ and let A_n decrease to $\emptyset$. Each A_n can be written

$$A_n = \cup_{1 \leq i \leq k_n} (x_i^n, y_i^n],$$

with $y_i^n < x_{i+1}^n$. Let $\varepsilon > 0$. By hypothesis (iii) there exists a number z such that $F(-z) \leq \varepsilon$ and $1 - F(z) \leq \varepsilon$. For each n, i there exists $a_i^n \in (x_i^n, y_i^n]$ such that $F(a_i^n) - F(x_i^n) \leq \frac{\varepsilon}{2^{i+n}}$, by (ii) (right continuity). Set

$$B_n' = \cup_{1 \leq i \leq k_n} \{(a_i^n, y_i^n] \cap (-z, z]\}, \qquad B_n = \cup_{m \leq n} B_m'.$$

Note that $B_n' \in \mathcal{B}_0$ and $B_n' \subset A_n$, hence $B_n \in \mathcal{B}_0$ and $B_n \subset A_n$. Furthermore, $A_n \backslash B_n \subset \cup_{m \leq n}(A_m \backslash B_m')$, hence

$$P(A_n) - P(B_n) \leq P((-z, z]^c) + \sum_{m=1}^{n} P((A_m \backslash B_m') \cap (-z, z])$$

$$\leq P((-z, z]^c) + \sum_{m=1}^{n} \sum_{i=1}^{k_n} P((x_i^n, a_i^n])$$

$$\leq F(-z) + 1 - F(z) + \sum_{m=1}^{n} \sum_{i=1}^{k_n} \{F(a_i^n) - F(x_i^n)\} \leq 3\varepsilon. \ (7.2)$$

Furthermore observe that $\overline{B_n} \subset A_n$ (where $\overline{B_n}$ is the closure of B_n), hence $\cap_{n=1}^{\infty} \overline{B_n} = \emptyset$ by hypothesis. Also $\overline{B_n} \subset [-z, z]$, hence each $\overline{B_n}$ is a compact set. It is a property of compact spaces[1] (known as "The Finite Intersection Property") that for closed sets F_β, $\cap_{\beta \in B} F_\beta \neq \emptyset$ if and only if $\cap_{\beta \in C} F_\beta \neq \emptyset$ for *all* finite subcollections C of B. Since in our case $\cap_{n=1}^{\infty} \overline{B_n} = \emptyset$, by the Finite Intersection Property we must have that there exists an m such that $\overline{B_n} = \phi$ for all $n \geq m$. Therefore $B_n = \phi$ for all $n \geq m$, hence $P(B_n) = 0$ for all $n \geq m$. Finally then

$$P(A_n) = P(A_n) - P(B_n) \leq 3\varepsilon$$

by (7.2), for all $n \geq m$. Since ε was arbitrary, we have $P(A_n) \downarrow 0$.

(Observe that this rather lengthy proof would become almost trivial if the sequence k_n above were bounded; but although A_n decreases to the empty set, it is not usually true). □

Corollary 7.1. *Let F be the distribution function of the probability P on* **R**. *Denoting by $F(x-)$ the left limit of F at x (which exists since F is nondecreasing), for all $x < y$ we have*

---

[1] For a definition of a *compact space* and the Finite intersection Property one can consult (for example) [12, p.81].

(i) $P((x, y]) = F(y) - F(x)$,
(ii) $P([x, y]) = F(y) - F(x-)$,
(iii) $P([x, y)) = F(y-) - F(x-)$,
(iv) $P((x, y)) = F(y-) - F(x)$,
(v) $P(\{x\}) = F(x) - F(x-)$,

and in particular $P(\{x\}) = 0$ for all x if and only the function F is continuous.

Proof. (i) has already been shown. For (ii) we write

$$P((x - \frac{1}{n}, y]) = F(y) - F(x - \frac{1}{n})$$

by (i). The left side converges to $F(y) - F(x-)$ as $n \to \infty$ by definition of the left limit of F; the right side converges to $P([x, y])$ by Theorem 2.3 because the sequence of intervals $(x - \frac{1}{n}, y]$ decreases to $[x, y]$. The claims (iii), (iv) and (v) are proved similarly. $\square$

Examples.
    We first consider two general examples:

1. If f is positive (and by "positive" we mean nonnegative; otherwise we say "strictly positive") and Riemann-integrable and $\int_{-\infty}^{\infty} f(x)dx = 1$, the function $F(x) = \int_{-\infty}^{x} f(y)dy$ is a distribution function of a probability on **R**; the function f is called its *density*. (It is *not* true that each distribution function admits a density, as the following example shows).
2. Let $\alpha \in$ **R**. A "point mass" probability on **R** (also known as "Dirac measure") is one that satisfies

$$P(A) = \begin{cases} 1 \text{ if } \alpha \in A, \\ 0 \text{ otherwise.} \end{cases}$$

Its distribution function is

$$F(x) = \begin{cases} 0 \text{ if } x < \alpha, \\ 1 \text{ if } x \geq \alpha. \end{cases}$$

This probability is also known as the *Dirac mass* at point α.

In the examples 3 through 10 below we define the distribution by its density function f; that is, we specify $f(x)$, and then the distribution function F corresponding to f is $F(x) = \int_{-\infty}^{x} f(u)du$. For f to be a density we need $f \geq 0$ and $\int_{-\infty}^{\infty} f(x)dx = 1$, which the reader can check is indeed the case for examples 3–10. We abuse language a bit by referring to the density f alone as the distribution, since it does indeed determine uniquely the distribution.

3. $f(x) = \begin{cases} \dfrac{1}{b-a} & \text{if } a \leq x \leq b, \\ 0 & \text{otherwise,} \end{cases}$

is called the *Uniform distribution on* $[a, b]$.

The uniform distribution is the continuous analog of the idea that "each point is equally likely"; this corresponds to a flat density function over the relevant interval $[a, b]$.

4. $f(x) = \begin{cases} \beta e^{-\beta x} & \text{if } x \geq 0, \\ 0 & \text{if } x < 0, \end{cases}$

is called the *Exponential distribution with parameter* $\beta > 0$.

The exponential distribution is often used to model the lifetime of objects whose decay has "no memory"; that is, if X is exponential, then the probability of an object lasting t more units of time given it has lasted s units already, is the same as the probability of a new object lasting t units of time. The lifetimes of light bulbs (for example) are often modeled this way; thus if one believes the model it is pointless to replace a working light bulb with a new one. This memoryless property characterizes the exponential distribution: see Exercises 9.20 and 9.21.

5. $f(x) = \begin{cases} \dfrac{\beta^\alpha}{\Gamma(\alpha)} x^{\alpha-1} e^{-\beta x} & x \geq 0, \\ 0 & x < 0, \end{cases}$

is called the *Gamma distribution with parameters* α, β $(0 < \alpha < \infty$ and $0 < \beta < \infty$; Γ denotes the gamma function) [2].

The Gamma distribution arises in various applications. One example is in reliability theory: if one has a part in a machine with an exponential (β) lifetime, one can build in reliability by including $n - 1$ back-up components. When a component fails, a back-up is used. The resulting lifetime then has a Gamma distribution with parameters (n, β). (See Exercise 15.17 in this regard.) The Gamma distribution also has a relationship to the Poisson distribution (see Exercise 9.22) as well as to the chi square distribution (see Example 6 in Chapter 15). The chi square distribution is important in Statistics: See the Remark at the end of Chapter 11.

6. $f(x) = \begin{cases} \alpha \beta^\alpha x^{\alpha-1} e^{-(\beta x)^\alpha} & \text{if } x \geq 0, \\ 0 & \text{if } x < 0, \end{cases}$

is called the *Weibull distribution with parameters* α, β $(0 < \alpha < \infty, 0 < \beta < \infty)$.

The Weibull distribution arises as a generalization of the exponential distribution for the modeling of lifetimes. This can be expressed in terms of its "hazard rate"; see for example Exercise 9.23.

7. $f(x) = \dfrac{1}{\sqrt{2\pi}\sigma} e^{-(x-\mu)^2/2\sigma^2}$ if $-\infty < x < \infty$ is called the *Normal distribution with parameters* (μ, σ^2), $(-\infty < \mu < \infty, 0 < \sigma^2 < \infty)$. It is also

---

[2] The Gamma function is defined to be $\Gamma(\alpha) = \int_0^\infty x^{\alpha-1} e^{-x} dx$, $\alpha > 0$; it follows from the definition that $\Gamma(\alpha) = (\alpha - 1)!$ for $\alpha \in \mathbf{N}$, and $\Gamma(\frac{1}{2}) = \sqrt{\pi}$.

known as the *Gaussian distribution*. Standard notation for the Normal with parameters μ and σ^2 is $N(\mu, \sigma^2)$.

We discuss the Normal Distribution at length in Chapters 16 and 21; it is certainly the most important distribution in probability and it is central to much of the subject of Statistics.

8. Let $g_{\mu,\sigma^2}(x) = \frac{1}{\sqrt{2\pi}\sigma} e^{-(x-\mu)^2/2\sigma^2}$, the normal density. Then

$$f(x) = \begin{cases} \dfrac{1}{x} g_{\mu,\sigma^2}(\log x) & \text{if } x > 0, \\ 0 & \text{if } x \leq 0, \end{cases}$$

is called the *Lognormal distribution with parameters* μ, $\sigma^2 (-\infty < \mu < \infty, 0 < \sigma^2 < \infty)$.

The lognormal distribution is used for numerous and varied applications to model nonnegative quantitative random phenomena. It is also known as the Galton–McAlister distribution and in Economics it is sometimes called the Cobb–Douglas distribution, where it has been used to model production data. It has been used to model drug dosage studies, lengths of words and sentences, lifetimes of mechanical systems, wildlife populations, and disease incubation periods.

9. $f(x) = \dfrac{\beta}{2} e^{-\beta|x-\alpha|}$ if $-\infty < x < \infty$ is called the *double exponential distribution with parameters* α, β ($-\infty < \alpha < \infty, 0 < \beta < \infty$). (It is also known as the *Laplace distribution*).

10. $f(x) = \dfrac{1}{\beta\pi} \dfrac{1}{1 + (x-\alpha)^2/\beta^2}$ if $-\infty < x < \infty$ is called the *Cauchy distribution with parameters* α, β ($-\infty < \alpha < \infty, 0 < \beta < \infty$).

The Cauchy distribution (named after Baron Louis–Augustin Cauchy (1789–1857)) is often used for counter-examples in Probability theory and was first proposed for that reason,[3] since it has very "heavy tails", which lead to the absence of nice properties. Nevertheless it is used in mechanics and electricity and in particular is useful for calibration problems in technical scientific fields.

---

[3] Indeed Poisson used it as early as 1824 to demonstrate a case where the Central Limit Theorem breaks down (the Central Limit Theorem is presented in Chapter 21). Later it was central in a large dispute between Cauchy and Bienaymé. It was this dispute that gave rise to its name as the Cauchy distribution.

Exercises for Chapters 6 and 7

7.1 Let $(A_n)_{n\geq 1}$ be any sequence of pairwise disjoint events and P a probability. Show that $\lim_{n\to\infty} P(A_n) = 0$.

7.2* Let $(A_\beta)_{\beta\in B}$ be a family of pairwise disjoint events. Show that if $P(A_\beta) > 0$, each $\beta \in B$, then B must be countable.

7.3 Show that the maximum of the Gamma density occurs at $x = \frac{\alpha-1}{\beta}$, for $\alpha \geq 1$.

7.4 Show that the maximum of the Weibull density occurs at $x = \frac{1}{\beta}(\frac{\alpha-1}{\alpha})^{\frac{1}{\alpha}}$, for $\alpha \geq 1$.

7.5 Show that the maximum of the Normal density occurs at $x = \mu$.

7.6 Show that the maximum of the Lognormal density occurs at $x = e^\mu e^{-\sigma^2}$.

7.7 Show that the maximum of the double exponential density occurs at $x = \alpha$.

7.8 Show that the Gamma and Weibull distributions both include the Exponential as a special case by taking $\alpha = 1$.

7.9 Show that the uniform, normal, double exponential, and Cauchy densities are all symmetric about their midpoints.

7.10 A distribution is called *unimodal* if the density has exactly one absolute maximum. Show that the normal, exponential, double exponential, Cauchy, Gamma, Weibull, and Lognormal are unimodal.

7.11 Let $P(A) = \int_{-\infty}^{\infty} 1_A(x)f(x)dx$ for a nonnegative function f with $\int_{-\infty}^{\infty} f(x)dx = 1$. Let $A = \{x_0\}$, a singleton (that is, the set A consists of one single point on the real line). Show that A is a Borel set and also a null set (that is, $P(A) = 0$).

7.12 Let P be as given in Exercise 7.11. Let B be a set with countable cardinality (that is, the number of points in B can be infinite, but only countably infinite). Show that B is a null set for P.

7.13 Let P and B be as given in Exercise 7.12. Suppose A is an event with $P(A) = \frac{1}{2}$. Show that $P(A \cup B) = \frac{1}{2}$ as well.

7.14 Let $A_1, \ldots, A_n, \ldots$ be a sequence of null sets. Show that $B = \cup_{i=1}^{\infty} A_i$ is also a null set.

7.15 Let X be a r.v. defined on a countable Probability space. Suppose $E\{|X|\} = 0$. Show that $X = 0$ except possibly on a null set. Is it possible to conclude, in general, that $X = 0$ everywhere (i.e., for all ω)? [Ans.: No]

7.16 * Let F be a distribution function. Show that in general F can have an infinite number of jump discontinuities, but that there can be at most countably many.

7.17 Suppose a distribution function F is given by

$$F(x) = \frac{1}{4}1_{[0,\infty)}(x) + \frac{1}{2}1_{[1,\infty)}(x) + \frac{1}{4}1_{[2,\infty)}(x).$$

Let P be given by

$$P((-\infty, x]) = F(x).$$

Then find the probabilities of the following events:

a) $A = (-\frac{1}{2}, \frac{1}{2})$
b) $B = (-\frac{1}{2}, \frac{3}{2})$
c) $C = (\frac{2}{3}, \frac{5}{2})$
d) $D = [0, 2)$
e) $E = (3, \infty)$

7.18 Suppose a function F is given by

$$F(x) = \sum_{i=1}^{\infty} \frac{1}{2^i} 1_{[\frac{1}{i}, \infty)}.$$

Show that it is the distribution function of a probability on **R**.

Let us define P by $P((-\infty, x]) = F(x)$. Find the probabilities of the following events:

a) $A = [1, \infty)$
b) $B = [\frac{1}{10}, \infty)$
c) $C = \{0\}$
d) $D = [0, \frac{1}{2})$
e) $E = (-\infty, 0)$
f) $G = (0, \infty)$

8 Random Variables

In Chapter 5 we considered random variables defined on a countable probability space $(\Omega, \mathcal{A}, P)$. We now wish to consider an arbitrary abstract space, countable or not. If X maps Ω into a state space $(F, \mathcal{F})$, then what we will often want to compute is the probability that X takes its values in a given subset of the state space. We take these subsets to be elements of the σ-algebra $\mathcal{F}$ of subsets of F. Thus, we will want to compute $P(\{\omega : X(\omega) \in A\}) = P(X \in A) = P(X^{-1}(A))$, which are three equivalent ways to write the same quantity. The third is enlightening: in order to compute $P(X^{-1}(A))$, we need $X^{-1}(A)$ to be an element of $\mathcal{A}$, the σ-algebra on Ω on which P is defined. This motivates the following definition.

Definition 8.1. (a) *Let $(E, \mathcal{E})$ and $(F, \mathcal{F})$ be two measurable spaces. A function $X : E \to F$ is called* measurable *(relative to $\mathcal{E}$ and $\mathcal{F}$) if $X^{-1}(\Lambda) \in \mathcal{E}$, for all $\Lambda \in \mathcal{F}$. (One also writes $X^{-1}(\mathcal{F}) \subset \mathcal{E}$.)*
(b) *When $(E, \mathcal{E}) = (\Omega, \mathcal{A})$, a measurable function X is called a random variable (r.v.).*
(c) *When $F = \mathbf{R}$, we usually take $\mathcal{F}$ to be the Borel σ-algebra $\mathcal{B}$ of $\mathbf{R}$. We will do this henceforth without special mention.*

Theorem 8.1. *Let $\mathcal{C}$ be a class of subsets of F such that $\sigma(\mathcal{C}) = \mathcal{F}$. In order for a function $X : E \to F$ to be measurable (w.r.t. the σ-algebras $\mathcal{E}$ and $\mathcal{F}$), it is necessary and sufficient that $X^{-1}(\mathcal{C}) \subset \mathcal{E}$.*

Proof. The necessity is clear, and we show sufficiency. That is, suppose that $X^{-1}(C) \in \mathcal{E}$ for all $C \in \mathcal{C}$. We need to show $X^{-1}(\Lambda) \in \mathcal{E}$ for all $\Lambda \in \mathcal{F}$. First note that $X^{-1}(\cup_n \Lambda_n) = \cup_n X^{-1}(\Lambda_n)$, $X^{-1}(\cap_n \Lambda_n) = \cap_n X^{-1}(\Lambda_n)$, and $X^{-1}(\Lambda^c) = (X^{-1}(\Lambda))^c$. Define $\mathcal{B} = \{A \in \mathcal{F} : X^{-1}(A) \in \mathcal{E}\}$. Then $\mathcal{C} \subset \mathcal{B}$, and since X^{-1} commutes with countable intersections, countable unions, and complements, we have that $\mathcal{B}$ is also a σ-algebra. Therefore $\mathcal{B} \supset \sigma(\mathcal{C})$, and also $\mathcal{F} \supset \mathcal{B}$, and since $\mathcal{F} = \sigma(\mathcal{C})$ we conclude $\mathcal{F} = \mathcal{B}$, and thus $X^{-1}(\mathcal{F}) \subset \sigma(X^{-1}(\mathcal{C})) \subset \mathcal{E}$. $\qquad\square$

We have seen that a probability measure P on $\mathbf{R}$ is characterized by the quantities $P((-\infty, a])$. Thus the distribution measure P^X on $\mathbf{R}$ of a random variable X should be characterized by $P^X((-\infty, a]) = P(X \le a)$ and what is perhaps surprisingly nice is that being a random variable is further characterized only by events of the form $\{\omega: X(\omega) \leq a\} = \{X \leq a\}$. Indeed, what this amounts to is that a function is measurable — and hence a random variable — if and only if its distribution function is defined.

Corollary 8.1. *Let $(F, \mathcal{F}) = (\mathbf{R}, \mathcal{B})$ and let $(E, \mathcal{E})$ be an arbitrary measurable space. Let X, X_n be real-valued functions on E.*

a) *X is measurable if and only if $\{X \leq a\} = \{\omega: X(\omega) \leq a\} = X^{-1}((-\infty, a])$ $\in \mathcal{E}$, for each a; or iff $\{X < a\} \in \mathcal{E}$, each $a \in \mathbf{R}$.*

b) *If X_n are measurable, $\sup X_n$, $\inf X_n$, $\limsup_{n \to \infty} X_n$ and $\liminf_{n \to \infty} X_n$ are all measurable.*

c) *If X_n are measurable and if X_n converges pointwise to X, then X is measurable.*

Proof.

(a) From Theorem 2.1, we know for the Borel sets $\mathcal{B}$ on $\mathbf{R}$ that $\mathcal{B} = \sigma(\mathcal{C})$ where $\mathcal{C} = \{(-\infty, a]; a \in \mathbf{R}\}$. By hypothesis $X^{-1}(\mathcal{C}) \subset \mathcal{E}$, so (a) follows from Theorem 8.1.

(b) Since X_n is measurable, $\{X_n \leq a\} \in \mathcal{E}$. Therefore $\{\sup_n X_n \leq a\} = \cap_n \{X_n \leq a\} \in \mathcal{E}$ for each a. Hence $\sup_n X_n$ is measurable by part (a). Analogously $\{\inf_n X_n < a\} = \cup_n \{X_n < a\} \in \mathcal{E}$ and thus $\inf_n X_n$ is measurable by part (a). Note further that $\limsup_{n \to \infty} X_n = \inf_n \sup_{m \geq n} X_m = \inf_n Y_n$, where $Y_n = \sup_{m \geq n} X_m$. We have just seen each Y_n is measurable, and we have also seen that $\inf_n Y_n$ is therefore measurable; hence $\limsup_{n \to \infty} X_n$ is measurable. Analogously $\liminf_{n \to \infty} X_n = \sup_n \inf_{m \geq n} X_m$ is measurable.

(c) If $\lim_{n \to \infty} X_n = X$, then $X = \limsup_{n \to \infty} X_n = \liminf_{n \to \infty} X_n$ (because the limit exists by hypothesis). Since $\limsup_{n \to \infty} X_n$ is measurable and equal to X, we conclude X is measurable as well.

$\square$

Theorem 8.2. *Let X be measurable from $(E, \mathcal{E})$ into $(F, \mathcal{F})$, and Y measurable from $(F, \mathcal{F})$ into $(G, \mathcal{G})$; then $Y \circ X$ is measurable from $(E, \mathcal{E})$ into $(G, \mathcal{G})$.*

Proof. Let $A \in \mathcal{G}$. Then $(Y \circ X)^{-1}(A) = X^{-1}(Y^{-1}(A))$. Since Y is measurable, $B = Y^{-1}(A) \in \mathcal{F}$. Since X is measurable, $X^{-1}(B) \in \mathcal{E}$. $\square$

A *topological space* is an abstract space with a collection of open sets;[1] the collection of open sets is called the *topology* of the space. An abstract definition of a *continuous function* is as follows: given two topological spaces $(E, \mathcal{U})$ and $(F, \mathcal{V})$ (where $\mathcal{U}$ are the open sets of E and $\mathcal{V}$ are the open sets of

---

[1] A "collection of open sets" is a collection of sets such that any union of sets in the collection is also in the collection, and any finite intersection of open sets in the collection is also in the collection.

$\mathcal{F}$), then a *continuous function* $f: E \to F$ is a function such that $f^{-1}(A) \in \mathcal{U}$ for each $A \in \mathcal{V}$. (This is written concisely as $f^{-1}(\mathcal{V}) \subset \mathcal{U}$.) The *Borel σ-algebra* of a topological space $(E, \mathcal{U})$ is $\mathcal{B} = \sigma(\mathcal{U})$. (The open sets do not form a σ-algebra by themselves: they are not closed under complements or under countable intersections.)

Theorem 8.3. *Let $(E, \mathcal{U})$ and $(F, \mathcal{V})$ be two topological spaces, and let $\mathcal{E}$, $\mathcal{F}$ be their Borel σ-algebras. Every continuous function X from E into F is then measurable (also called "Borel").*

Proof. Since $\mathcal{F} = \sigma(\mathcal{V})$, by Theorem 8.1 it suffices to show that $X^{-1}(\mathcal{V}) \subset \mathcal{E}$. But for $O \in \mathcal{V}$, we know $X^{-1}(O)$ is open and therefore in $\mathcal{E}$, as $\mathcal{E}$ being the Borel σ-algebra, it contains the class $\mathcal{U}$ of open sets of E. $\square$

Recall that for a subset A of E, the *indicator function* $1_A(x)$ is defined to be

$$1_A(x) = \begin{cases} 1 \text{ if } x \in A, \\ 0 \text{ if } x \notin A \end{cases}.$$

Thus the function $1_A(x)$, usually written 1_A with the argument x being implicit, "indicates" whether or not a given x is in A. (Sometimes the function 1_A is known as the "characteristic function of A" and it is also written χ_A; this terminology and notation is somewhat out of date.)

Theorem 8.4. *Let $(F, \mathcal{F}) = (\mathbf{R}, \mathcal{B})$, and $(E, \mathcal{E})$ be any measurable space.*

a) *An indicator 1_A on E is measurable if and only if $A \in \mathcal{E}$.*
b) *If $X_1, \ldots, X_n$ are real-valued measurable functions on $(E, \mathcal{E})$, and if f is Borel on $\mathbf{R}^n$, then $f(X_1, \ldots, X_n)$ is measurable.*
c) *If X, Y are measurable, so also are $X + Y$, XY, $X \vee Y$ (a short-hand for $\max(X, Y)$), $X \wedge Y$ (a short-hand for $\min(X, Y)$), and X/Y (if $Y \neq 0$).*

Proof. (a) If $B \subset \mathbf{R}$, we have

$$(1_A)^{-1}(B) = \begin{cases} \emptyset & \text{if } 0 \notin B, \ 1 \notin B \\ A & \text{if } 0 \notin B, \ 1 \in B \\ A^c & \text{if } 0 \in B, \ 1 \notin B \\ E & \text{if } 0 \in B, \ 1 \in B \end{cases}$$

The result follows.

(b) The Borel σ-algebra $\mathcal{B}^n$ on $\mathbf{R}^n$ is generated by the quadrants $\prod_{i \leq n}(-\infty, a_i]$, by the exact same proof as was used in Theorem 2.1. Let X denote the vector-valued function $X = (X_1, \ldots, X_n)$. Thus $X: E \to \mathbf{R}^n$. Then $X^{-1}(\prod_{i \leq n}(-\infty, a_i]) = \cap_{i \leq n}\{X_i \leq a_i\} \in \mathcal{E}$, and therefore X is a measurable function from $(E, \mathcal{E})$ into $(\mathbf{R}^n, \mathcal{B}^n)$. The statement (b) then follows from Theorem 8.2.

(c) Note that the function $f_1: \mathbf{R}^2 \to \mathbf{R}$ given by $f_1(x, y) = x + y$ is continuous. So also are $f_2(x, y) = xy$; $f_3(x, y) = x \vee y = \max(x, y)$; $f_4(x, y) =$

$x \wedge y = \min(x,y)$. The function $f_5(x,y) = \frac{x}{y}$ is continuous from $\mathbf{R} \times (\mathbf{R} \setminus \{0\})$ into $\mathbf{R}$. Therefore (c) follows from (b) together with Theorem 8.3 (that continuous functions are Borel measurable). □

If X is a r.v. on $(\Omega, \mathcal{A}, P)$, with values in $(E, \mathcal{E})$, then the *distribution measure* (or *law*) of X is defined by

$$P^X(B) = P(X^{-1}(B)) = (P \circ X^{-1})(B) = P(\omega \colon X(\omega) \in B) = P(X \in B),$$

$\forall B \in \mathcal{E}$. All four different ways of writing $P^X(B)$ on the right side above are used in mathematics, but the most common is

$$P^X(B) = P(X \in B)$$

where the "ω" is not explicitly written but rather implicitly understood to be there. This allows us to avoid specifying the space Ω (which is often difficult to construct mathematically), and simply work with Probability measures on $(E, \mathcal{E})$. Sometimes P^X is also called the *image* of P by X.

Since X^{-1} commutes with taking unions and intersections and since $X^{-1}(E) = \Omega$, we have

Theorem 8.5. *The distribution of X, (or the law of X), is a probability measure on $(E, \mathcal{E})$.*

When X is a real-valued r.v., its distribution P^X is a probability on $\mathbf{R}$, which is entirely characterized by its distribution function:

$$F_X(x) = P^X((-\infty, x]) = P(X \leq x).$$

The function F_X is called the *cumulative distribution function of the r.v. X*. When F_X admits a density f_X (i.e. $F_X(x) = \int_{-\infty}^x f_X(y)dy$ for all $x \in \mathbf{R}$), we also say that the function f_X is the *probability density of the r.v. X*. Often the cumulative distribution function is referred to as a "cdf", or simply a "distribution function." Analogously the probability density function is referred to as a "pdf", or simply a "density."

9 Integration with Respect to a Probability Measure

Let $(\Omega, \mathcal{A}, P)$ be a probability space. We want to define the *expectation*, or what is equivalent, the "integral", of general random variables. We have of course already done this for random variables defined on a countable space Ω. The general case (for arbitrary Ω) is more delicate.

Let us begin with special cases.

Definition 9.1. a) *A r.v. X is called* simple *if it takes on only a finite number of values and hence can be written in the form*

$$X = \sum_{i=1}^{n} a_i 1_{Ai}, \tag{9.1}$$

where $a_i \in \mathbf{R}$, and $A_i \in \mathcal{A}$, $1 \le i \le n$ (Such an X is clearly measurable; conversely if X is measurable and takes on the values $a_1, \ldots, a_n$ it must have the representation (9.1) with $A_i = \{X = a_i\}$; a simple r.v. has of course many different representations of the form (9.1).)

b) *If X is simple, its* expectation *(or "integral" with respect to P) is the number*

$$E\{X\} = \sum_{i=1}^{n} a_i P(A_i). \tag{9.2}$$

(This is also written $\int X(\omega) P(d\omega)$ and even more simply $\int X dP$.)

A little algebra shows that $E\{X\}$ does not depend on the particular representation (9.1) chosen for X.

Let X, Y be two simple random variables and β a real number. We clearly can write both X and Y in the form (9.1), with the same subsets A_i which form a partition of Ω, and with numbers a_i for X and b_i for Y. Then βX and $X + Y$ are again in the form (9.1) with the same A_i and with the respective numbers βa_i and $a_i + b_i$. Thus $E\{\beta X\} = \beta E\{X\}$ and $E\{X + Y\} = E\{X\} + E\{Y\}$; that is expectation is linear on the vector space of all simple r.v.'s. If further $X \le Y$ we have $a_i \le b_i$ for all i, and thus $E\{X\} \le E\{Y\}$.

Next we define expectation for positive random variables. For X positive (by this, we assume that X may take all values in $[0, \infty]$, including $+\infty$: this innocuous extension is necessary for the coherence of some of our further results), let

$$E\{X\} = \sup(E\{Y\} : Y \text{ a simple r.v. with } 0 \le Y \le X). \qquad (9.3)$$

This supremum always exists in $[0, \infty]$. Since expectation is a positive operator on the set of simple r.v.'s, it is clear that the definition above for $E\{X\}$ coincides with Definition 9.1.

Note that $E\{X\} \ge 0$, but we can have $E\{X\} = \infty$, even when X is never equal to $+\infty$.

Finally let X be an arbitrary r.v. Let $X^+ = \max(X, 0)$ and $X^- = -\min(X, 0)$. Then $X = X^+ - X^-$, and X^+, X^- are positive r.v.'s. Note that $|X| = X^+ + X^-$.

Definition 9.2. (a) *A r.v. X has a finite expectation (is "integrable") if both $E\{X^+\}$ and $E\{X^-\}$ are finite. In this case, its expectation is the number*

$$E\{X\} = E\{X^+\} - E\{X^-\}, \qquad (9.4)$$

also written $\int X(\omega) dP(\omega)$ or $\int X dP$. (If $X \ge 0$ then $X^- = 0$ and $X^+ = X$ and, since obviously $E\{0\} = 0$, this definition coincides with (9.3)).

We write $\mathcal{L}^1$ to denote the set of all integrable random variables. (Sometimes we write $\mathcal{L}^1(\Omega, \mathcal{A}, P)$ to remove any possible ambiguity.)

(b) *A r.v. X admits an expectation if $E\{X^+\}$ and $E\{X^-\}$ are not both equal to $+\infty$. Then the expectation of X is still given by (9.4), with the conventions $+\infty + a = +\infty$ and $-\infty + a = -\infty$ when $a \in \mathbf{R}$. (If $X \ge 0$ this definition again coincides with (9.3); note that if X admits an expectation, then $E\{X\} \in [-\infty, +\infty]$, and X is integrable if and only if its expectation is finite.)*

Remark 9.1. When Ω is finite or countable we have thus two different definitions for the expectation of a r.v. X, the one above and the one given in Chapter 5. In fact these two definitions coincide: it is enough to verify this for a simple r.v. X, and in this case the formulas (5.1) and (9.2) are identical.

The next theorem contains the most important properties of the expectation operator. The proofs of (d), (e) and (f) are considered hard and could be skipped.

Theorem 9.1. (a) *$\mathcal{L}^1$ is a vector space, and expectation is a linear map on $\mathcal{L}^1$, and it is also positive (i.e., $X \ge 0 \Rightarrow E\{X\} \ge 0$). If further $0 \le X \le Y$ are two r.v. and $Y \in \mathcal{L}^1$, then $X \in \mathcal{L}^1$ and $E\{X\} \le E\{Y\}$.*
(b) *$X \in \mathcal{L}^1$ iff $|X| \in \mathcal{L}^1$ and in this case $|E\{X\}| \le E\{|X|\}$. In particular any bounded r.v. is integrable.*
(c) *If $X = Y$ almost surely[1] (a.s.), then $E\{X\} = E\{Y\}$.*
(d) *(Monotone convergence theorem): If the r.v.'s X_n are positive and increasing a.s. to X, then $\lim_{n \to \infty} E\{X_n\} = E\{X\}$ (even if $E\{X\} = \infty$).*

---

[1] $X = Y$ a.s. if $P(X = Y) = P(\{\omega : X(\omega) = Y(\omega)\}) = 1$

(e) *(Fatou's lemma): If the r.v.'s X_n satisfy $X_n \geq Y$ a.s. ($Y \in \mathcal{L}^1$), all n, we have $E\{\liminf_{n \to \infty} X_n\} \leq \liminf_{n \to \infty} E\{X_n\}$. In particular if $X_n \geq 0$ a.s. all n, then $E\{\liminf_{n \to \infty} X_n\} \leq \liminf_{n \to \infty} E\{X_n\}$.*

(f) *(Lebesgue's dominated convergence theorem): If the r.v.'s X_n converge a.s. to X and if $|X_n| \leq Y$ a.s. $\in \mathcal{L}^1$, all n, then $X_n \in \mathcal{L}^1$, $X \in \mathcal{L}^1$, and $E\{X_n\} \to E\{X\}$.*

The a.s. equality between random variables is clearly an equivalence relation, and two equivalent (i.e. almost surely equal) random variables have the same expectation: thus one can define a space L^1 by considering "$\mathcal{L}^1$ modulo this equivalence relation". In other words, an element of L^1 is an equivalence class, that is a collection of all r.v. in $\mathcal{L}^1$ which are pairwise a.s. equal. In view of (c) above, one may speak of the "expectation" of this equivalence class (which is the expectation of any one element belonging to this class). Since further the addition of random variables or the product of a r.v. by a constant preserve a.s. equality, the set L^1 is also a vector space. Therefore we commit the (innocuous) abuse of identifying a r.v. with its equivalence class, and commonly write $X \in L^1$ instead of $X \in \mathcal{L}^1$.

If $1 \leq p < \infty$, we define $\mathcal{L}^p$ to be the space of r.v.'s such that $|X|^p \in \mathcal{L}^1$; L^p is defined analogously to L^1. That is, L^p is $\mathcal{L}^p$ modulo the equivalence relation "almost surely". Put more simply, two elements of $\mathcal{L}^p$ that are a.s. equal are considered to be representatives of one element of L^p. We will use in this book only the spaces L^1 and L^2 (that is $p = 1$ or 2).

Before proceeding to the proof of Theorem 9.1 itself, we show two auxiliary results.

Result 1. *For every positive r.v. X there exists a sequence $(X_n)_{n \geq 1}$ of positive simple r.v.'s which increases toward X as n increases to infinity.* An example of such a sequence is given by

$$X_n(\omega) = \begin{cases} k2^{-n} & \text{if } k2^{-n} \leq X(\omega) < (k+1)2^{-n} \text{ and } 0 \leq k \leq n2^n - 1, \\ n & \text{if } X(\omega) \geq n. \end{cases}$$

$$(9.5)$$

Result 2. *If X is a positive r.v., and if $(X_n)_{n \geq 1}$ is any sequence of positive simple r.v.'s increasing to X, then $E\{X_n\}$ increases to $E\{X\}$.*

To see this, observe first that the sequence $E\{X_n\}$ increases to a limit a, which satisfies $a \leq E\{X\}$ by (9.3). To obtain that indeed $a = E\{X\}$, and in view of (9.3) again, it is clearly enough to prove that if Y is a simple r.v. such that $0 \leq Y \leq X$, then $E\{Y\} \leq a$.

The variable Y takes on m different values, say $a_1, \ldots, a_m$, and set $A_k = \{Y = a_k\}$. Choose $\varepsilon \in (0, 1]$. The r.v. $Y_{n,\varepsilon} = (1 - \varepsilon)Y 1_{\{(1-\varepsilon)Y \leq X_n\}}$ takes the value $(1 - \varepsilon)a_k$ on the set $A_{k,n,\varepsilon} = A_k \cap \{(1 - \varepsilon)Y \leq X_n\}$ and 0 on the set $\{(1 - \varepsilon)Y > X_n\}$. Furthermore it is obvious that $Y_{n,\varepsilon} \leq X_n$, hence using (9.2), we obtain

$$E\{Y_{n,\varepsilon}\} = (1 - \varepsilon) \sum_{k=1}^{m} a_k P(A_{k,n,\varepsilon}) \le E\{X_n\}. \qquad (9.6)$$

Now recall that $Y \le \lim_n X_n$, hence $(1 - \varepsilon)Y < \lim_n X_n$ as soon as $Y > 0$, hence clearly $A_{k,n,\varepsilon} \to A_k$ as $n \to \infty$. An application of Theorem 2.4 yields $P(A_{k,n,\varepsilon}) \to P(A_k)$, hence taking the limit in (9.6) gives

$$(1 - \varepsilon) \sum_{k=1}^{m} a_k \, P(A_k) = (1 - \varepsilon)E\{Y\} \le a.$$

Letting $\varepsilon \to 0$ in the above, we deduce that $E\{Y\} \le a$, hence our Result 2.

Proof. (a) Let X, Y be two nonnegative r.v.'s and $\alpha \in \mathbf{R}_+$. By Result 1 above we associate with X, Y simple r.v.'s X_n, Y_n increasing to X, Y respectively. Then αX_n and $X_n + Y_n$ are also simple r.v.'s increasing respectively to αX and $X + Y$. Using that expectation is linear over simple r.v.'s and Result 2 above, we readily obtain $E\{\alpha X\} = \alpha E\{X\}$ and $E\{X+Y\} = E\{X\}+E\{Y\}$. If further $X \le Y$, that $E\{X\} \le E\{Y\}$ readily follows from (9.3).

From this we first deduce the last two claims of (a). Since for (possibly negative) r.v.'s X and Y and $\alpha \in \mathbf{R}$ we have that $(\alpha X)^+ + (\alpha X)^- \le |\alpha|(X^+ + X^-)$ and $(X + Y)^+ + (X + Y)^- \le X^+ + X^- + Y^+ + Y^-$, we deduce next that $\mathcal{L}^1$ is a vector space. Finally since $E\{X\} = E\{X^+\} - E\{X^-\}$ we deduce that expectation is linear.

(b) If $X \in \mathcal{L}^1$, then $E\{X^+\} < \infty$ and $E\{X^-\} < \infty$. Since $|x| = x^+ + x^-$, we have $E\{|X|\} = E\{X^+\} + E\{X^-\} < \infty$ as well, so $|X| \in \mathcal{L}^1$. Conversely if $E\{|X|\} < \infty$, then $E\{X^+ + X^-\} < \infty$, and since $E\{X^+ + X^-\} = E\{X^+\} + E\{X^-\}$ and both terms are nonnegative, we have that they are also both finite and $X \in \mathcal{L}^1$.

(c) Suppose $X = Y$ a.s. and assume first $X \ge 0$, $Y \ge 0$. Let $A = \{\omega : X(\omega) \ne Y(\omega)\} = \{X \ne Y\}$. Then $P(A) = 0$. Also,

$$E\{Y\} = E\{Y1_A + Y1_{A^c}\} = E\{Y1_A\} + E\{Y1_{A^c}\} = E\{Y1_A\} + E\{X1_{A^c}\}.$$

Let Y_n be simple and Y_n increase to Y. Then $Y_n 1_A$ are simple and $Y_n 1_A$ increase to $Y1_A$ too. Since Y_n is simple it is bounded, say by N. Then

$$0 \le E\{Y_n 1_A\} \le E\{N1_A\} = NP(A) = 0.$$

Therefore $E\{Y1_A\} = 0$. Analogously, $E\{X1_A\} = 0$. Finally we have

$$E\{Y\} = E\{Y1_A\} + E\{X1_{A^c}\} = 0 + E\{X1_{A^c}\} = E\{X1_{A^c}\} + E\{X1_A\}$$
$$= E\{X\}.$$

We conclude by noting that if $Y = X$ a.s., then also $Y^+ = X^+$ a.s. and $Y^- = X^-$ a.s., and (c) follows.

(d) For each fixed n choose an increasing sequence $Y_{n,k}$, $k = 1, 2, 3, \ldots$ of positive simple r.v.'s increasing to X_n (Result 1), and set

$$Z_k = \max_{n \le k} Y_{n,k}.$$

Then $(Z_k)_{k \ge 1}$ is a non-decreasing sequence of positive simple r.v.'s, and thus it has a limit $Z = \lim_{k \to \infty} Z_k$. Also

$$Y_{n,k} \le Z_k \le X_k \le X \quad \text{a.s.} \quad \text{for } n \le k$$

which implies that

$$X_n \le Z \le X \quad \text{a.s.}$$

Next if we let $n \to \infty$ we have $Z = X$ a.s. Since expectation is a positive operator we have

$$E\{Y_{n,k}\} \le E\{Z_k\} \le E\{X_k\} \qquad \text{for } n \le k$$

Fix n and let $k \to \infty$. Using Result 2, we obtain

$$E\{X_n\} \le E\{Z\} \le \lim_{k \to \infty} E\{X_k\}.$$

Now let $n \to \infty$ to obtain:

$$\lim_{n \to \infty} E\{X_n\} \le E\{Z\} \le \lim_{k \to \infty} E\{X_k\},$$

and since the left and right sides are the same, they must equal the middle; by (c) and $X = Z$ a.s., we deduce the result.

(e) Note that we could replace X_n with $\hat{X}_n = X_n - Y$ and then $\hat{X}_n \ge 0$, $\hat{X}_n \in \mathcal{L}^1$, and $E\{\liminf_{n \to \infty} \hat{X}_n\} \le \liminf_{n \to \infty} E\{\hat{X}_n\}$ if and only if $E\{\liminf_{n \to \infty} X_n\} \le \liminf_{n \to \infty} E\{X_n\}$, because

$$\liminf_{n \to \infty} \hat{X}_n = (\liminf_{n \to \infty} X_n) - Y.$$

Therefore without loss of generality we assume $X_n \ge 0$ a.s., each n.

Set $Y_n = \inf_{k \ge n} X_k$. Then Y_n are also random variables and form a non-decreasing sequence. Moreover

$$\lim_{n \to \infty} Y_n = \liminf_{n \to \infty} X_n.$$

Since $X_n \ge Y_n$, we have $E\{X_n\} \ge E\{Y_n\}$, whence

$$\liminf_{n \to \infty} E\{X_n\} \ge \lim_{n \to \infty} E\{Y_n\} = E\{\lim_{n \to \infty} Y_n\} = E\{\liminf_{n \to \infty} X_n\}$$

by the Monotone Convergence Theorem (part (d) of this theorem).

(f) Set $U = \liminf_{n \to \infty} X_n$ and $V = \limsup_{n \to \infty} X_n$. By hypothesis $U = V = X$ a.s. We also have $|X_n| \le Y$ a.s., hence $|X| \le Y$ as well, hence X_n and X are integrable. On the one hand $X_n \ge -Y$ a.s. and $-Y \in \mathcal{L}^1$, so Fatou's lemma (e) yields

$$E\{U\} \leq \liminf_{n\to\infty} E\{X_n\}.$$

We also have $-X_n \geq -Y$ a.s. and $-V = \liminf_{n\to\infty} -X_n$, so another application of Fatou's lemma yields

$$-E\{V\} = E\{-V\} \geq \liminf_{n\to\infty} E\{-X_n\} = -\limsup_{n\to\infty} E\{X_n\}.$$

Putting together these two inequalities and applying (c) yields

$$E\{X\} = E\{U\} \leq \liminf_{n\to\infty} E\{X_n\} \leq \limsup_{n\to\infty} E\{X_n\} \leq E\{V\} = E\{X\}.$$

This completes the proof.   □

A useful consequence of Lebesgue's Dominated Convergence Theorem (Theorem 9.1(f)) is the next result which allows us to interchange summation and expectation. Since an infinite series is a limit of partial sums and an expectation is also a limit, the interchange of expectation and summation amounts to changing the order of taking two limits.

Theorem 9.2. *Let X_n be a sequence of random variables.*

(a) *If the X_n's are all positive, then*

$$E\left\{\sum_{n=1}^{\infty} X_n\right\} = \sum_{n=1}^{\infty} E\{X_n\}, \tag{9.7}$$

both sides being simultaneously finite or infinite.

(b) *If $\sum_{n=1}^{\infty} E\{|X_n|\} < \infty$, then $\sum_{n=1}^{\infty} X_n$ converges a.s. and the sum of this series is integrable and moreover (9.7) holds.*

Proof. Let $S_n = \sum_{k=1}^{n} |X_k|$ and $T_n = \sum_{k=1}^{n} X_k$. Then

$$E\{S_n\} = E\left\{\sum_{k=1}^{n} |X_k|\right\} = \sum_{k=1}^{n} E\{|X_k|\},$$

and the sequence S_n clearly increases to the limit $S = \sum_{k=1}^{\infty} |X_k|$ (which may be finite for some values of ω and infinite for others). Therefore by the Monotone Convergence Theorem (Theorem 9.1(d)) we have:

$$E\{S\} = \lim_{n\to\infty} E\{S_n\} = \sum_{k=1}^{\infty} E\{|X_k|\} < \infty.$$

If all X_n's are positive, then $S_n = T_n$ and this proves (a). If the X_n's are not necessarily positive, but $\sum_{n=1}^{\infty} E\{|X_n|\} < \infty$, we deduce also that $E\{S\} < \infty$.

Now, for every $\varepsilon > 0$ we have $1_{\{S=\infty\}} \leq \varepsilon S$, hence

$$P(S = \infty) = E\{1_{\{S=\infty\}}\} \leq \varepsilon E\{S\}.$$

Then $E\{S\} < \infty$ and since the choice of ε is arbitrary, we have that $P(S = \infty) = 0$: we deduce that $\sum_{k=1}^{\infty} X_k$ is an absolutely convergent series a.s. and its sum, say T, is the limit of the sequence T_n. Moreover

$$|T_n| \leq S_n \leq S$$

and S is in L^1. Hence by the Dominated Convergence Theorem (Theorem 9.1(f)) we have that

$$E\left\{\sum_{k=1}^{\infty} X_k\right\} = E\{\lim_{n\to\infty} T_n\} = E\{T\},$$

which is (9.7). □

Recall that L^1 and L^2 are the sets of equivalence classes of integrable (resp. square-integrable) random variables for the a.s. equivalence relation.

Theorem 9.3. a) *If $X, Y \in L^2$, we have $XY \in L^1$ and the* Cauchy-Schwarz *inequality:*

$$|E\{XY\}| \leq \sqrt{E\{X^2\}E\{Y^2\}};$$

b) *We have $L^2 \subset L^1$, and if $X \in L^2$, then $E\{X\}^2 \leq E\{X^2\}$;*
c) *The space L^2 is a linear space, i.e. if $X, Y \in L^2$ and $\alpha, \beta \in \mathbf{R}$, then $\alpha X + \beta Y \in L^2$ (we will see in Chapter 22 that in fact L^2 is a Hilbert space).*

Proof. (a) We have $|XY| \leq X^2/2 + Y^2/2$, hence $X, Y \in L^2$ implies $XY \in L^1$. For every $x \in \mathbf{R}$ we have

$$0 \leq E\{(xX + Y)^2\} = x^2 E\{X^2\} + 2x E\{XY\} + E\{Y^2\}. \qquad (9.8)$$

The discriminant of the quadratic equation in x given in (9.8) is

$$\sqrt{4\{(E\{XY\})^2 - E\{X^2\}E\{Y^2\}\}},$$

and since the equation is always nonnegative,

$$E\{XY\}^2 - E\{X^2\}E\{Y^2\} \leq 0,$$

which gives the Cauchy-Schwarz inequality.
(b) Let $X \in L^2$. Since $X = X \cdot 1$ and since the function equal to 1 identically obviously belongs to L^2 with $E\{1^2\} = 1$, the claim follows readily from (a).
(c) Let $X, Y \in L^2$. Then for constants α, β, $(\alpha X + \beta Y)^2 \leq 2\alpha^2 X^2 + 2\beta^2 Y^2$ is integrable and $\alpha X + \beta Y \in L^2$ and L^2 is a vector space. □

If $X \in L^2$, the *variance* of X, written $\sigma^2(X)$ or σ_X^2, is

$$\text{Var}(X) = \sigma^2(X) \equiv E\{(X - E\{X\})^2\}.$$

(Note that $X \in L^2 \Rightarrow X \in L^1$, so $E\{X\}$ exists.) Let $\mu = E\{X\}$. Then $\text{Var}(X) =$

$$
\begin{aligned}
E\{(X - \mu)^2\} &= E\{X^2\} - 2\mu E\{X\} + \mu^2 \\
&= E\{X^2\} - 2\mu^2 + \mu^2 \\
&= E\{X^2\} - \mu^2.
\end{aligned}
$$

Thus we have as well the trivial but nonetheless very useful equality:

$$\sigma^2(X) = E\{X^2\} - E\{X\}^2.$$

Theorem 9.4 (Chebyshev's Inequality).

$$P\{|X| \geq a\} \leq \frac{E\{X^2\}}{a^2}.$$

Proof. Since $a^2 1_{\{|X| \geq a\}} \leq X^2$, we have $E\{a^2 1_{\{|X| \geq a\}}\} \leq E\{X^2\}$, or

$$a^2 P(|X| \geq a) \leq E\{X^2\};$$

and dividing by a^2 gives the result.   □

Chebyshev's inequality is also known as the Bienaymé-Chebyshev inequality, and often is written equivalently as

$$P\{|X - E\{X\}| \geq a\} \leq \frac{\sigma^2(X)}{a^2},$$

The next theorem is useful; both Theorem 9.5 and Corollary 9.1 we call the Expectation Rule, as they are vital tools for calculating expectations. It shows in particular that the expectation of a r.v. depends only on its distribution.

Theorem 9.5 (Expectation Rule). *Let X be a r.v. on $(\Omega, \mathcal{A}, P)$, with values in $(E, \mathcal{E})$, and distribution P^X. Let $h \colon (E, \mathcal{E}) \to (\mathbf{R}, \mathcal{B})$ be measurable.*

a) *We have $h(X) \in \mathcal{L}^1(\Omega, \mathcal{A}, P)$ if and only if $h \in \mathcal{L}^1(E, \mathcal{E}, P^X)$.*
b) *If either h is positive, or if it satisfies the equivalent conditions in (a), we have:*

$$E\{h(X)\} = \int h(x) P^X(dx). \tag{9.9}$$

Proof. Recall that the distribution measure P^X is defined by $P^X(B) = P(X^{-1}(B))$. Therefore

$$E\{1_B(X)\} = P(X^{-1}(B)) = P^X(B) = \int 1_B(x)P^X(dx).$$

Thus if h is simple, (9.9) holds by the above and linearity. If h is positive, let h_n be simple, positive and increase to h. Then

$$
\begin{aligned}
E\{h(X)\} &= E\{\lim_{n\to\infty} h_n(X)\} \\
&= \lim_{n\to\infty} E\{h_n(X)\} \\
&= \lim_{n\to\infty} \int h_n(x)P^X(dx) \\
&= \int \lim_{n\to\infty} h_n(x)P^X(dx) \\
&= \int h(x)P^X(dx)
\end{aligned}
$$

where we have used the Monotone Convergence Theorem twice. This proves (b) when h is positive, and applied to $|h|$ it also proves (a) (recalling that a r.v. belongs to $\mathcal{L}^1$ if and only if the expectation of its absolute value is finite).

If h is not positive, we write $h = h^+ - h^-$ and deduce the result by subtraction. □

The next result can be proved as a consequence of Theorem 9.5, but we prove it in Chapter 11 (Corollary 11.1) so we omit its proof here.

Corollary 9.1 (Expectation Rule). *Suppose X is a random variable that has a density f. (That is, $F(x) = P(X \leq x)$ and $F(x) = \int_{-\infty}^{x} f(u)du, -\infty < x < \infty$.) If $E\{|h(X)|\} < \infty$ or if h is positive, then $E\{h(X)\} = \int h(x)f(x)dx$.*

Examples:

1. Let X be exponential with parameter α. Then

$$E\{h(X)\} = \int_0^\infty h(x)\alpha e^{-\alpha x}dx.$$

In particular, if $h(x) = x$, we have

$$E\{X\} = \int_0^\infty \alpha x e^{-\alpha x}dx = \frac{1}{\alpha},$$

by integration by parts. Thus the mean of an exponential random variable is $1/\alpha$.

2. Let X be normal (or Gaussian) with parameters (μ, σ^2). Then $E\{X\} = \mu$, since

$$E\{X\} = \int_{-\infty}^\infty \frac{1}{\sqrt{2\pi}\sigma} x e^{-(x-\mu)^2/2\sigma^2}dx.$$

To see this, let $y = x - \mu$; then $x = y + \mu$, and

$$E\{X\} = \int_{-\infty}^{\infty} \frac{1}{\sqrt{2\pi}\sigma}(y + \mu)e^{-y^2/2\sigma^2}\,dy$$

$$= \int_{-\infty}^{\infty} \frac{1}{\sqrt{2\pi}\sigma}ye^{-y^2/2\sigma^2}\,dy + \mu\int_{-\infty}^{\infty} \frac{1}{\sqrt{2\pi}\sigma}e^{-y^2/2\sigma^2}\,dy.$$

The first integral is the integral of an odd function, so it is zero; the second is equal to $\mu \int_{-\infty}^{\infty} f(x)dx = \mu \cdot 1 = \mu$.

3. Let X be Cauchy with density function

$$f(x) = \frac{1}{\pi}\frac{1}{1+x^2}.$$

We have $E\{X^+\} = E\{X^-\} = E\{|X|\} = \infty$ and the mean $E\{X\}$ of a Cauchy random variable does not exist. Indeed

$$E\{X^+\} = \int_0^{\infty} \frac{x}{\pi(1+x^2)}\,dx + \int_{-\infty}^0 \frac{0}{\pi(1+x^2)}\,dx$$

$$\geq \frac{1}{\pi}\int_1^{\infty} \frac{1}{2x}\,dx = \infty$$

since $\frac{x}{1+x^2} \geq 0$ for all $x \geq 0$ and $\frac{x}{1+x^2} \geq \frac{1}{2x}$ for $x > 1$. That $E\{X^-\} = \infty$ is proved similarly.

Exercises for Chapters 8 and 9

9.1 Let $X : (\Omega, \mathcal{A}) \to (\mathbf{R}, \mathcal{B})$ be a r.v. Let

$$\mathcal{F} = \{A : A = X^{-1}(B), \text{ some } B \in \mathcal{B}\} = X^{-1}(\mathcal{B}).$$

Show that X is measurable as a function from $(\Omega, \mathcal{F})$ to $(\mathbf{R}, \mathcal{B})$.

9.2 * Let $(\Omega, \mathcal{A}, P)$ be a probability space, and let $\mathcal{F}$ and $\mathcal{G}$ be two σ-algebras on Ω. Suppose $\mathcal{F} \subset \mathcal{A}$ and $\mathcal{G} \subset \mathcal{A}$ (we say in this case that $\mathcal{F}$ and $\mathcal{G}$ are *sub σ-algebras* of $\mathcal{A}$). The σ-algebras $\mathcal{F}$ and $\mathcal{G}$ are *independent* if for any $A \in \mathcal{F}$, any $B \in \mathcal{G}$, $P(A \cap B) = P(A)P(B)$. Suppose $\mathcal{F}$ and $\mathcal{G}$ are independent, and a r.v. X is measurable from both $(\Omega, \mathcal{F})$ to $(\mathbf{R}, \mathcal{B})$ and from $(\Omega, \mathcal{G})$ to $(\mathbf{R}, \mathcal{B})$. Show that X is a.s. constant; that is, $P(X = c) = 1$ for some constant c.

9.3 * Given $(\Omega, \mathcal{A}, P)$, let $\mathcal{A}' = \{A \cup N : A \in \mathcal{A}, N \in \mathcal{N}\}$, where $\mathcal{N}$ are the null sets (as in Theorem 6.4). Suppose $X = Y$ a.s. where X and Y are two real-valued functions on Ω. Show that $X : (\Omega, \mathcal{A}') \to (\mathbf{R}, \mathcal{B})$ is measurable if and only if $Y : (\Omega, \mathcal{A}') \to (\mathbf{R}, \mathcal{B})$ is measurable.

9.4 * Let $X \in \mathcal{L}^1$ on $(\Omega, \mathcal{A}, P)$ and let A_n be a sequence of events such that $\lim_{n \to \infty} P(A_n) = 0$. Show that $\lim_{n \to \infty} E\{X 1_{A_n}\} = 0$. (Caution: We are not assuming that $\lim_{n \to \infty} X 1_{A_n} = 0$ a.s.)

9.5 * Given $(\Omega, \mathcal{A}, P)$, suppose X is a r.v. with $X \geq 0$ a.s. and $E\{X\} = 1$. Define $Q : \mathcal{A} \to \mathbf{R}$ by $Q(A) = E\{X 1_A\}$. Show that Q defines a probability measure on $(\Omega, \mathcal{A})$.

9.6 For Q as in Exercise 9.5, show that if $P(A) = 0$, then $Q(A) = 0$. Give an example that shows that $Q(A) = 0$ does not in general imply $P(A) = 0$.

9.7 * For Q as in Exercise 9.5, suppose also $P(X > 0) = 1$. Let E_Q denote expectation with respect to Q. Show that $E_Q\{Y\} = E_P\{YX\}$.

9.8 Let Q be as in Exercise 9.5, and suppose that $P(X > 0) = 1$.

(a) Show that $\frac{1}{X}$ is integrable for Q.
(b) Define $R : \mathcal{A} \to \mathbf{R}$ by $R(A) = E_Q\{\frac{1}{X} 1_A\}$. Show that R is exactly the probability measure P of Exercise 9.5. (*Hint:* Use Exercise 9.7.)

9.9 Let Q be as in Exercise 9.8. Show that $Q(A) = 0$ implies $P(A) = 0$ (compare with Exercise 9.6).

9.10 Let X be uniform over (a, b). Show that $E\{X\} = \frac{a+b}{2}$.

9.11 Let X be an integrable r.v. with density $f(x)$, and let $\mu = E\{X\}$. Show that

$$\text{Var}(X) = \sigma^2(X) = \int_{-\infty}^{\infty} (x - \mu)^2 f(x) dx.$$

9.12 Let X be uniform over (a, b). Show that $\sigma^2(X) = \frac{(b-a)^2}{12}$.

9.13 Let X be Cauchy with density $\frac{1}{\pi(1+(x-a)^2)}$. Show that $\sigma^2(X)$ is not defined, and $E\{X^2\} = \infty$.

9.14 The *beta function* is $B(r, s) = \frac{\Gamma(r)\Gamma(s)}{\Gamma(r+s)}$, where Γ is the gamma function. Equivalently

$$B(r, s) = \int_0^1 t^{r-1}(1-t)^{s-1}dt \quad (r > 0, s > 0).$$

X is said to have a *beta distribution* if the density f of its distribution measure is

$$f(x) = \begin{cases} \dfrac{x^{r-1}(1-x)^{s-1}}{B(r, s)} & \text{if } 0 \le x \le 1, \\ 0 & \text{if } x < 0 \text{ or } x > 1. \end{cases}$$

Show that for X having a beta distribution with parameters (r, s) $(r > 0, s > 0)$, then

$$E\{X^k\} = \frac{B(r+k, s)}{B(r, s)} = \frac{\Gamma(r+k)\Gamma(r+s)}{\Gamma(r)\Gamma(r+s+k)},$$

for $k \ge 0$. Deduce that

$$E\{X\} = \frac{r}{r+s},$$

$$\sigma^2(X) = \frac{rs}{(r+s)^2(r+s+1)}.$$

The beta distribution is a rich family of distributions on the interval $[0, 1]$. It is often used to model random proportions.

9.15 Let X have a lognormal distribution with parameters (μ, σ^2). Show that

$$E\{X^r\} = e^{r\mu + \frac{1}{2}\sigma^2 r^2}$$

and deduce that $E\{X\} = e^{\mu + \frac{1}{2}\sigma^2}$ and $\sigma_X^2 = e^{2\mu + \sigma^2}(e^{\sigma^2} - 1)$. (*Hint:* $E\{X^r\} = \int_0^\infty x^r f(x)dx$ where f is the lognormal density; make the change of variables $y = \log(x) - \mu$ to obtain

$$E\{X^r\} = \int_{-\infty}^\infty \frac{1}{\sqrt{2\pi\sigma^2}} e^{(r\mu + ry - y^2/2\sigma^2)} dy.)$$

9.16 The gamma distribution is often simplified to a one parameter distribution. A r.v. X is said to have the *standard gamma distribution* with parameter α if the density of its distribution measure is given by

$$f(x) = \begin{cases} \dfrac{x^{\alpha-1}e^{-x}}{\Gamma(\alpha)} & \text{if } x \ge 0, \\ 0 & \text{if } x < 0. \end{cases}$$

That is, $\beta = 1$. (Recall $\Gamma(\alpha) = \int_0^\infty t^{\alpha-1}e^{-t}dt$.) Show that for X standard gamma with parameter α, then

$$E\{X^k\} = \frac{\Gamma(\alpha + k)}{\Gamma(\alpha)} \quad (k \geq 0).$$

Deduce that X has mean α and also variance α.

9.17 * Let X be a nonnegative r.v. with mean μ and variance σ^2, both finite. Show that for any $b > 0$,

$$P\{X \geq \mu + b\sigma\} \leq \frac{1}{1 + b^2}.$$

(*Hint:* Consider the function $g(x) = \frac{\{(x-\mu)b+\sigma\}^2}{\sigma^2(1+b^2)^2}$ and that $E\{((X - \mu)b + \sigma)^2\} = \sigma^2(b^2 + 1)$.)

9.18 Let X be a r.v. with mean μ and variance σ^2, both finite. Show that

$$P\{\mu - d\sigma < X < \mu + d\sigma\} \geq 1 - \frac{1}{d^2}.$$

(Note that this is interesting only for $d > 1$.)

9.19 Let X be normal (or Gaussian) with parameters $\mu = 0$ and $\sigma^2 = 1$. Show that $P(X > x) \leq \frac{1}{x\sqrt{2\pi}} e^{-\frac{1}{2}x^2}$, for $x > 0$.

9.20 Let X be an exponential r.v.. Show that $P\{X > s + t \mid X > s\} = P\{X > t\}$ for $s > 0$, $t > 0$. This is known as the "memoryless property" of the exponential.

9.21 * Let X be a r.v. with the property that $P\{X > s+t \mid X > s\} = P\{X > t\}$. Show that if $h(t) = P\{X > t\}$, then h satisfies *Cauchy's equation*:

$$h(s + t) = h(s)h(t) \quad (s > 0, t > 0)$$

and show that X is exponentially distributed (*Hint*: use the fact that h is continuous from the right, so Cauchy's equation can be solved).

9.22 Let α be an integer and suppose X has distribution Gamma (α, β). Show that $P(X \leq x) = P(Y \geq \alpha)$, where Y is Poisson with parameter $\lambda = x\beta$. (Hint: Recall $\Gamma(\alpha) = (\alpha - 1)!$ and write down $P(X \leq x)$, and then use integration by parts with $u = t^{\alpha-1}$ and $dv = e^{-t/\beta}dt$.)

9.23 The *Hazard Rate* of a nonnegative random variable X is defined by

$$h_X(t) = \lim_{\varepsilon \to 0} \frac{P(t \leq X < t + \varepsilon \mid X \geq t)}{\varepsilon}$$

when the limit exists. The hazard rate can be thought of as the probability that an object does not survive an infinitesimal amount of time after time t. The memoryless property of the exponential gives rise to a constant rate. A Weibull random variable can be used as well to model lifetimes. Show that:

a) If X is exponential (λ), then its hazard rate is $h_X(t) = \lambda$;

b) If X is Weibull (α, β), then its hazard rate is $h_X(t) = \alpha \beta^\alpha t^{\alpha-1}$.

9.24 A positive random variable X has the *logistic distribution* if its distribution function is given by

$$F(x) = P(X \le x) = \frac{1}{1 + e^{-(x-\mu)/\beta}}; \quad (x > 0),$$

for parameters (μ, β), $\beta > 0$.

a) Show that if $\mu = 0$ and $\beta = 1$, then a density for X is given by

$$f(x) = \frac{e^{-x}}{(1 + e^{-x})^2};$$

b) Show that if X has a logistic distribution with parameters (μ, β), then X has a hazard rate and it is given by $h_X(t) = \left(\frac{1}{\beta}\right) F(t)$.

10 Independent Random Variables

Recall that two events A and B are independent if knowledge that B has occurred does not change the probability that A will occur: that is, $P(A \mid B) = P(A)$. This of course is algebraically equivalent to the statement $P(A \cap B) = P(A)P(B)$. The latter expression generalizes easily to a finite number of events: $A_1, \ldots, A_n$ are independent if $P(\cap_{i=J} A_i) = \prod_{i=J} P(A_i)$, for every subset J of $\{1, \ldots, n\}$ (see Definition 3.1).

For two random variables X and Y to be independent we want knowledge of Y to leave unchanged the probabilities that X will take on certain values, which roughly speaking means that the events $\{X \in A\}$ and $\{Y \in B\}$ are independent for any choice of A and B in the σ-algebras of the state space of X and Y. This is more easily expressed in terms of the σ-algebras generated by X and Y: Recall that if $X: (\Omega, \mathcal{A}) \to (E, \mathcal{E})$, then $X^{-1}(\mathcal{E})$ is a sub σ-algebra of $\mathcal{A}$, called the *σ-algebra generated by* X. This σ-algebra is often denoted $\sigma(X)$.

Definition 10.1. a) *Sub σ-algebras $(\mathcal{A}_i)_{i \in I}$ of $\mathcal{A}$, are independent if for every finite subset J of I, and all $A_i \in \mathcal{A}_i$, one has*

$$P\left(\cap_{i \in J} A_i\right) = \prod_{i \in J} P(A_i).$$

b) *Random variables $(X_i)_{i \in I}$, with values in $(E_i, \mathcal{E}_i)$, are independent if the generated σ-algebras $X_i^{-1}(\mathcal{E}_i)$ are independent.*

We will next, for notational simplicity, consider only *pairs* (X, Y) of random variables. However the results extend without difficulty to finite families of r.v.'s.

Note that X and Y are not required to take values in the same space: X can take its values in $(E, \mathcal{E})$ and Y in $(F, \mathcal{F})$.

Theorem 10.1. *In order for X and Y to be independent, it is necessary and sufficient to have any one of the following conditions holding:*

a) $P(X \in A, Y \in B) = P(X \in A)P(Y \in B)$ *for all $A \in \mathcal{E}$, $B \in \mathcal{F}$;*
b) $P(X \in A, Y \in B) = P(X \in A)P(Y \in B)$ *for all $A \in \mathcal{C}$, $B \in \mathcal{D}$, where $\mathcal{C}$ and $\mathcal{D}$ are respectively classes of sets stable under finite intersections which generate $\mathcal{E}$ and $\mathcal{F}$;*

c) $f(X)$ and $g(Y)$ are independent for each pair (f, g) of measurable functions;

d) $E\{f(X)g(Y)\} = E\{f(X)\}E\{g(Y)\}$ for each pair (f, g) of functions bounded measurable, or positive measurable.

e) Let E and F be metric spaces and let $\mathcal{E}$, $\mathcal{F}$ be their Borel σ-algebras. Then $E\{f(X)g(Y)\} = E\{f(X)\}E\{g(Y)\}$ for each pair (f, g) of bounded, continuous functions.

Proof. (a) This is a restatement of the definition, since $X^{-1}(\mathcal{E})$ is exactly all events of the form $\{X \in A\}$, for $A \in \mathcal{E}$. (a)$\Rightarrow$(b): This is trivial since $\mathcal{C} \subset \mathcal{E}$ and $\mathcal{D} \subset \mathcal{F}$.

(a)$\Rightarrow$(b): This is evident.

(b)$\Rightarrow$(a): The collection of sets $A \in \mathcal{E}$ that verifies $P(X \in A, Y \in B) = P(X \in A)P(Y \in B)$ for a given $B \in \mathcal{D}$ is closed under increasing limits and by difference and it contains the class $\mathcal{C}$ by hypothesis, and this class $\mathcal{C}$ is closed by intersection. So the Monotone Class Theorem 6.2 yields that this collection is in fact $\mathcal{E}$ itself. In other words, Assumption (b) is satisfied with $\mathcal{C} = \mathcal{E}$. Then analogously by fixing $A \in \mathcal{E}$ and letting $\mathcal{J} = \{B \in \mathcal{F}\colon P(X \in A, Y \in B) = P(X \in A)P(Y \in B)\}$, we have $\mathcal{J} \supset \sigma(\mathcal{D})$ and thus $\mathcal{J} = \mathcal{F}$.

(c)$\Rightarrow$(a): We need only to take $f(x) = x$ and $g(y) = y$.

(a)$\Rightarrow$(c): Given f and g, note that

$$f(X)^{-1}(\mathcal{E}) = X^{-1}(f^{-1}(\mathcal{E})) \subset X^{-1}(\mathcal{E}).$$

Also, $g(Y)^{-1}(\mathcal{F}) \subset Y^{-1}(\mathcal{F})$, and since $X^{-1}(\mathcal{E})$ and $Y^{-1}(\mathcal{F})$ are independent, the two sub σ-algebras $f(X)^{-1}(\mathcal{E})$ and $g(Y)^{-1}(\mathcal{F})$ will also be.

(d)$\Rightarrow$(a): Take $f(x) = 1_A(x)$ and $g(y) = 1_B(y)$.

(a)$\Rightarrow$(d): We have (d) holds for indicator functions, and thus for simple functions (i.e., $f(x) = \sum_{i=1}^{k} a_i 1_{A_i}(x)$) by linearity. If f and g are positive, let f_n and g_n be simple positive functions increasing to f and g respectively. Observe that the products $f_n(X)g_n(Y)$ increase to $f(X)g(Y)$. Then

$$E\{f(X)g(Y)\} = E\left\{\lim_{n\to\infty} f_n(X)g_n(Y)\right\} = \lim_{n\to\infty} E\{f_n(X)g_n(Y)\}$$
$$= \lim_{n\to\infty} E\{f_n(X)\}E\{g_n(Y)\} = E\{f(X)\}E\{g(Y)\}$$

by the monotone convergence theorem. This gives the result when f and g are positive. When f and g are bounded we write $f = f^+ - f^-$ and $g = g^+ - g^-$ and we conclude by linearity.

(d)$\Rightarrow$(e): This is evident.

(e)$\Rightarrow$(b): It is enough to prove (b) when $\mathcal{C}$ and $\mathcal{D}$ are the classes of all closed sets of E and F (these classes are stable by intersection). Let for example A be a closed subset of E. If $d(x, A)$ denotes the distance between the point x and the set A, then $f_n(x) = \min(1, nd(x, A))$ is continuous, it satisfies $0 \le f_n \le 1$, and the sequence $(1 - f_n)$ decreases to the indicator function 1_A. Similarly with B a closed subset of F we associate continuous functions g_n decreasing to 1_B and having $0 \leq g_n \leq 1$. Then it suffices to reproduce the proof of the implication (a)$\Rightarrow$(d), substituting the monotone convergence theorem for the dominated convergence theorem.   □

Example: Let E and F be finite or countable. For the couple (X, Y) let

$$
\begin{aligned}
P_{ij}^{XY} &= P(X = i, Y = j) \\
&= P(\{\omega : X(\omega) = i \text{ and } Y(\omega) = j\}) \\
&= P\{(X = i) \cap (Y = j)\}.
\end{aligned}
$$

Then X and Y are independent if and only if $P_{ij}^{XY} = P_i^X P_j^Y$, as a consequence of Theorem 10.1.

We present more examples in Chapter 12.

We now wish to discuss "jointly measurable" functions. In general, if $\mathcal{E}$ and $\mathcal{F}$ are each σ-algebras on spaces E and F respectively, then the Cartesian product $\mathcal{E} \times \mathcal{F} = \{A \subset E \times F : A = \Lambda \times \Gamma, \Lambda \in \mathcal{E} \text{ and } \Gamma \in \mathcal{F}\}$ is *not* a σ-algebra on $E \times F$. Consequently we write $\sigma(\mathcal{E} \times \mathcal{F})$ to denote the smallest σ-algebra on $E \times F$ generated by $\mathcal{E} \times \mathcal{F}$. Such a construct is common, and we give it a special notation:

$$
\mathcal{E} \otimes \mathcal{F} = \sigma(\mathcal{E} \times \mathcal{F}).
$$

Theorem 10.2. *Let f be measurable: $(E \times F, \mathcal{E} \otimes \mathcal{F}) \to (\mathbf{R}, \mathcal{R})$. For each $x \in E$ (resp. $y \in F$), the "section" $y \to f(x, y)$ (resp. $x \to f(x, y)$) is an $\mathcal{F}$-measurable (resp. $\mathcal{E}$-measurable) function.*

Note: The converse to Theorem 10.2 is false in general.

Proof. First assume f is of the form $f(x, y) = 1_C(x, y)$, for $C \in \mathcal{E} \otimes \mathcal{F}$. Let $\mathcal{H} = \{C \in \mathcal{E} \otimes \mathcal{F} : y \to 1_C(x, y) \text{ is } \mathcal{F}\text{-measurable for each fixed } x \in E\}$. Then $\mathcal{H}$ is a σ-algebra and $\mathcal{H}$ contains $\mathcal{E} \times \mathcal{F}$, hence $\sigma(\mathcal{E} \times \mathcal{F}) \subset \mathcal{H}$. But by construction $\mathcal{H} \subset \sigma(\mathcal{E} \times \mathcal{F})$, so we have $\mathcal{H} = \mathcal{E} \otimes \mathcal{F}$. Thus we have the result for indicators and hence also for simple functions by linearity. If f is positive, let f_n be simple functions increasing to f. Then $g_n(y) = f_n(x, y)$ for x fixed is $\mathcal{F}$-measurable for each n, and since

$$
g(y) = \lim_{n \to \infty} g_n(y) = f(x, y),
$$

and since the limit of measurable functions is measurable, we have the result for f. Finally if f is arbitrary, take $f = f^+ - f^-$, and since the result holds for f^+ and f^-, it holds as well for f because the difference of two measurable functions is measurable.   □

Theorem 10.3 (Tonelli-Fubini). *Let P and Q be two probabilities on $(E, \mathcal{E})$ and $(F, \mathcal{F})$ respectively.*

a) *Define $R(A \times B) = P(A)Q(B)$, for $A \in \mathcal{E}$ and $B \in \mathcal{F}$. Then R extends uniquely to a probability on $(E \times F, \mathcal{E} \otimes \mathcal{F})$, written $P \otimes Q$.*

b) *For each function f that is $\mathcal{E} \otimes \mathcal{F}$-measurable, positive, or integrable with respect to $P \otimes Q$, the function $x \to \int f(x,y)Q(dy)$ is $\mathcal{E}$-measurable, the function $y \to \int f(x,y)P(dx)$ is $\mathcal{F}$-measurable and*

$$\int f\, dP \otimes Q = \int \left\{ \int f(x,y)Q(dy) \right\} P(dx)$$
$$= \int \left\{ \int f(x,y)P(dx) \right\} Q(dy).$$

Proof. (a) Let $C \in \mathcal{E} \otimes \mathcal{F}$, and let us write $C(x) = \{y \colon (x,y) \in C\}$. If $C = A \times B$, we have in this case $C(x) = B$ if $x \in A$ and $C(x) = \emptyset$ otherwise, hence:

$$R(C) = P \otimes Q(C) = P(A)Q(B) = \int P(dx)Q[C(x)].$$

Let $\mathcal{H} = \{C \in \mathcal{E} \otimes \mathcal{F} \colon x \to Q[C(x)] \text{ is } \mathcal{E}\text{-measurable}\}$. Then $\mathcal{H}$ is closed under increasing limit and differences, while $\mathcal{E} \times \mathcal{F} \subset \mathcal{H} \subset \mathcal{E} \otimes \mathcal{F}$, whence $\mathcal{H} = \mathcal{E} \otimes \mathcal{F}$ by the monotone class theorem. For each $C \in \mathcal{H} = \mathcal{E} \otimes \mathcal{F}$, we can now define (since $Q[C(x)]$ is measurable and positive)

$$R(C) = \int P(dx)Q[C(x)].$$

We need to show R is a probability measure. We have

$$R(\Omega) = R(E \times F) = \int_E P(dx)Q[F] = 1.$$

Let $C_n \in \mathcal{E} \otimes \mathcal{F}$ be pairwise disjoint and set $C = \cup_{n=1}^{\infty} C_n$. Then since the $C_n(x)$ also are pairwise disjoint and since Q is a probability measure, $Q[C(x)] = \sum_{n=1}^{\infty} Q[C_n(x)]$. Apply Theorem 9.2 to the probability measure P and to the functions $f_n(x) = Q[C_n(x)]$, to obtain

$$\sum_{n=1}^{\infty} R(C_n) = \sum_{n=1}^{\infty} \int f_n dP$$
$$= \int (\sum_{n=1}^{\infty} f_n) dP$$
$$= \int P(dx)Q[C(x)] = R(C).$$

Thus R is a probability measure. The uniqueness of R follows from Corollary 6.1.

(b) Note that we have already established part (b) in our proof of (a) for functions f of the form $f(x,y) = 1_C(x,y)$, $C \in \mathcal{E} \otimes \mathcal{F}$. The result follows for positive simple functions by linearity. If f is positive, $\mathcal{E} \otimes \mathcal{F}$-measurable, let f_n be simple functions increasing to f. Then

$$E_R(f) = \lim_{n\to\infty} E_R(f_n) = \lim_{n\to\infty} \int \left\{ \int f_n(x,y)Q(dy) \right\} P(dx).$$

But $x \to \int f_n(x,y)Q(dy)$ are functions that increase to $x \to \int f(x,y)Q(dy)$, hence by the monotone convergence theorem

$$= \int \left\{ \lim_{n\to\infty} \int f_n(x,y)Q(dy) \right\} P(dx),$$

and again by monotone convergence

$$= \int \left\{ \int \lim_{n\to\infty} f_n(x,y)Q(dy) \right\} P(dx) = \int \left\{ \int f(x,y)Q(dy) \right\} P(dx).$$

An analogous argument gives

$$= \int \left\{ \int f(x,y)P(dx) \right\} Q(dy).$$

Finally for general f it suffices to take $f = f^+ - f^-$ and the result follows.
$\square$

Corollary 10.1. *Let X and Y be two r.v. on $(\Omega, \mathcal{A}, P)$, with values in $(E, \mathcal{E})$ and $(F, \mathcal{F})$ respectively. The pair $Z = (X,Y)$ is a r.v. with values in $(E \times F, \mathcal{E} \otimes \mathcal{F})$, and the r.v.'s X, Y are independent if and only if the distribution $P^{(X,Y)}$ of the couple (X,Y) equals the product $P^X \otimes P^Y$ of the distributions of X and Y.*

Proof. Since $Z^{-1}(A \times B) = X^{-1}(A) \cap Y^{-1}(B)$ belongs to $\mathcal{A}$ as soon as $A \in \mathcal{E}$ and $B \in \mathcal{F}$, the measurability of Z follows from the definition of the product σ-algebra $\mathcal{E} \otimes \mathcal{F}$ and from Theorem 8.1.

X and Y are independent iff for all $A \in \mathcal{E}$ and $B \in \mathcal{F}$, we have

$$P\left((X,Y) \in A \times B\right) = P(X \in A)P(Y \in B),$$

or equivalently

$$P^{(X,Y)}(A \times B) = P^X(A)P^Y(B).$$

This is equivalent to saying that $P^{(X,Y)}(A \times B) = (P^X \otimes P^Y)(A \times B)$ for all $A \times B \in \mathcal{E} \otimes \mathcal{F})$, which by the uniqueness in Fubini's theorem is in turn equivalent to the fact that $P^{(X,Y)} = P^X \otimes P^Y$ on $\mathcal{E} \otimes \mathcal{F}$. $\square$

We digress slightly to discuss the *construction of a model with independent random variables*. Let μ be a probability measure on $(E, \mathcal{E})$. It is easy to construct a r.v. X, with values in E, whose distribution measure is μ: simply take $\Omega = E$; $\mathcal{A} = \mathcal{E}$; $P = \mu$; and let X be the identity: $X(x) = x$.

Slightly more complicated is the *construction of two independent random variables*, X and Y, with values in $(E, \mathcal{E})$, $(F, \mathcal{F})$, and given distribution measures μ and ν. We can do this as follows: take $\Omega = E \times F$; $\mathcal{A} = \mathcal{E} \otimes \mathcal{F}$; $P = \mu \otimes \nu$, and $X(x, y) = x$; $Y(x, y) = y$, where $(x, y) \in E \times F$.

Significantly more complicated, but very important for applications, is to *construct an infinite sequence of independent random variables* of given distributions. Specifically, for each n let X_n be defined on $(\Omega_n, \mathcal{A}_n P_n)$, and let us set

$$\Omega = \prod_{n=1}^{\infty} \Omega_n \qquad \text{(countable Cartesian product)}$$

$$\mathcal{A} = \bigotimes_{n=1}^{\infty} \mathcal{A}_n$$

where $\otimes_{n=1}^{\infty} \mathcal{A}_n$ denotes the smallest σ-algebra on Ω generated by all sets of the form

$$A_1 \times A_2 \times \ldots \times A_k \times \Omega_{k+1} \times \Omega_{k+2} \times \ldots, \qquad A_i \in \mathcal{A}_i; \qquad k = 1, 2, 3, \ldots .$$

That is, $\mathcal{A}$ is the smallest σ-algebra generated by finite Cartesian products of sets from the coordinate σ-algebras.

The next theorem is from general measure theory, and can be considered a (non trivial) extension of Fubini's theorem. We state it without proof.

Theorem 10.4. *Given $(\Omega_n, \mathcal{A}_n, P_n)$ probability spaces and $\Omega = \prod_{n=1}^{\infty} \Omega_n$, $\mathcal{A} = \otimes_{n=1}^{\infty} \mathcal{A}_n$, then there exists a probability P on $(\Omega, \mathcal{A})$, and it is unique, such that*

$$P(A_1 \times A_2 \times \ldots \times A_k \times \Omega_{k+1} \times \Omega_{k+2} \times \ldots) = \prod_{i=1}^{k} P_i(A_i)$$

for all $k = 1, 2, \ldots$ and $A_i \in \mathcal{A}_i$.

For X_n defined on $(\Omega_n, \mathcal{A}_n, P_n)$ as in Theorem 10.4, let $\widetilde{X}_n$ denote its *natural extension* to Ω as follows: for $\omega \in \Omega$, let $\omega = (\omega_1, \omega_2, \ldots, \omega_n, \ldots)$ with $\omega_i \in \Omega_i$, each i. Then

$$\widetilde{X}_n(\omega) = X_n(\omega_n).$$

Corollary 10.2. *Let X_n be defined on $(\Omega_n, \mathcal{A}_n, P_n)$, each n, and let $\widetilde{X}_n$ be its natural extension to $(\Omega, \mathcal{A}, P)$ as given above. Then $(\widetilde{X}_n)_{n \geq 1}$ are all independent, and the law of $\widetilde{X}_n$ on $(\Omega, \mathcal{A}, P)$ is identical to the law of X_n on $(\Omega_n, \mathcal{A}_n, P_n)$.*

Proof. We have

$$\widetilde{X}_n^{-1}(B_n) = \Omega_1 \times \ldots \times \Omega_{n-1} \times X_n^{-1}(B_n) \times \Omega_{n+1} \times \Omega_{n+2} \times \ldots,$$

and by Theorem 10.4 we have for $k = 1, 2, \ldots$:

$$P\left(\cap_{n=1}^{k}X_n^{-1}(B_n)\right) = P\left(X_1^{-1}(B_1) \times \ldots X_k^{-1}(B_k) \times \Omega_{k+1} \times \ldots\right)$$
$$= \prod_{n=1}^{k} P_n(X_n \in B_n),$$

and the result follows.   □

Next we wish to discuss some significant properties of independence. Let A_n be a sequence of events in $\mathcal{A}$. We define:

$$\limsup_{n\to\infty} A_n = \cap_{n=1}^{\infty}\left(\cup_{m\geq n}A_m\right) = \lim_{n\to\infty}\left(\cup_{m\geq n}A_m\right).$$

This event can be interpreted probabilistically as:

$$\limsup_{n\to\infty} A_n = \text{``}A_n \text{ occurs infinitely often''},$$

which means that A_n occurs for an infinite number of n. This is often abbreviated "i.o.", and thus we have:

$$\limsup_{n} A_n = \{A_n \text{ i.o.}\}.$$

Theorem 10.5 (Borel-Cantelli). *Let A_n be a sequence of events in $(\Omega, \mathcal{A}, P)$.*

a) *If $\sum_{n=1}^{\infty} P(A_n) < \infty$, then $P(A_n \text{ i.o.}) = 0$.*
b) *If $P(A_n \text{ i.o.}) = 0$ and if the A_n's are mutually independent, then $\sum_{n=1}^{\infty} P(A_n) < \infty$.*

Note: An alternative statement to (b) is: if A_n are mutually independent events, and if $\sum_{n=1}^{\infty} P(A_n) = \infty$, then $P(A_n \text{ i.o.}) = 1$. Hence for mutually independent events A_n, and since the sum $\sum_{n} P(A_n)$ has to be either finite or infinite, the event $\{A_n \text{ i.o.}\}$ has probability either 0 or 1; this is a particular case of the so-called zero-one law to be seen below.

Proof. (a) Let $a_n = P(A_n) = E\{1_{A_n}\}$. By Theorem 9.2(b) $\sum_{n=1}^{\infty} a_n < \infty$ implies $\sum_{n=1}^{\infty} 1_{A_n} < \infty$ a.s. On the other hand, $\sum_{n=1}^{\infty} 1_{A_n}(\omega) = \infty$ if and only if $\omega \in \limsup_{n\to\infty} A_n$. Thus we have (a).

(b) Suppose now the A_n's are mutually independent. Then

$$P(\limsup_{n\to\infty} A_n) = \lim_{n\to\infty} \lim_{k\to\infty} P\left(\cup_{m=n}^{k}A_m\right)$$
$$= \lim_{n\to\infty} \lim_{k\to\infty} \left(1 - P\left(\cap_{m=n}^{k}A_m^c\right)\right)$$
$$= 1 - \lim_{n\to\infty} \lim_{k\to\infty} \left(\prod_{m=n}^{k}(1 - P(A_m))\right)$$

by independence;

$$= 1 - \lim_{n \to \infty} \lim_{k \to \infty} \prod_{m=n}^{k} (1 - a_m)$$

where $a_m = P(A_m)$. By hypothesis $P(\limsup_{n \to \infty} A_n) = 0$, so $\lim_{n \to \infty} \lim_{k \to \infty} \prod_{m=n}^{k} (1 - a_m) = 1$. Therefore by taking logarithms we have

$$\lim_{n \to \infty} \lim_{k \to \infty} \sum_{m=n}^{k} \log(1 - a_m) = 0,$$

or

$$\lim_{n \to \infty} \sum_{m \geq n} \log(1 - a_m) = 0,$$

which means that $\sum_m \log(1 - a_m)$ is a convergent series. Since $|\log(1-x)| \geq x$ for $0 < x < 1$, we have that $\sum_m a_m$ is convergent as well. $\qquad \square$

Let now X_n be r.v.'s all defined on $(\Omega, \mathcal{A}, P)$. Define the σ-algebras

$$\mathcal{B}_n = \sigma(X_n)$$
$$\mathcal{C}_n = \sigma\left(\cup_{p \geq n} \mathcal{B}_p\right)$$
$$\mathcal{C}_\infty = \cap_{n=1}^{\infty} \mathcal{C}_n$$

$\mathcal{C}_\infty$ is called the *tail σ-algebra*.

Theorem 10.6 (Kolmogorov's Zero-one law). *Let X_n be independent r.v.'s, all defined on $(\Omega, \mathcal{A}, P)$, and let $\mathcal{C}_\infty$ be the corresponding tail σ-algebra. If $C \in \mathcal{C}_\infty$, then $P(C) = 0$ or 1.*

Proof. Let $\mathcal{D}_n = \sigma(\cup_{p<n} \mathcal{B}_p)$. By the hypothesis, $\mathcal{C}_n$ and $\mathcal{D}_n$ are independent, hence if $A \in \mathcal{C}_n$, $B \in \mathcal{D}_n$, then

$$P(A \cap B) = P(A)P(B). \qquad (10.1)$$

If $A \in \mathcal{C}_\infty$ we hence have (10.1) for all $B \in \cup \mathcal{D}_n$, hence also for all $B \in \mathcal{D} = \sigma(\cup \mathcal{D}_n)$, by the Monotone Class Theorem (Theorem 6.2). However $\mathcal{C}_\infty \subset \mathcal{D}$, whence we have (10.1) for $B = A \in \mathcal{C}_\infty$, which implies $P(A) = P(A)P(A) = P(A)^2$, hence $P(A) = 0$ or 1. $\qquad \square$

Consequences:

1. $\{\omega : \lim_{n \to \infty} X_n(\omega) \text{ exists}\} \in \mathcal{C}_\infty$, therefore X_n either converges a.s. or it diverges a.s.
2. Each r.v. which is $\mathcal{C}_\infty$ measurable is a.s. constant. In particular,

$$\limsup_{n \to \infty} X_n, \qquad \liminf_{n \to \infty} X_n,$$

$$\limsup_{n \to \infty} \frac{1}{n} \sum_{p \leq n} X_p, \qquad \liminf_{n \to \infty} \frac{1}{n} \sum_{p \leq n} X_p$$

are all a.s. constant. (Recall we are still assuming that X_n is a sequence of independent r.v.'s)

Exercises for Chapter 10

10.1 Let $f = (f_1, f_2): \Omega \to E \times F$. Show that $f:(\Omega, \mathcal{A}) \to (E \times F, \mathcal{E} \otimes \mathcal{F})$ is measurable if and only if f_1 is measurable from $(\Omega, \mathcal{A})$ to $(E, \mathcal{E})$ and f_2 is measurable from $(\Omega, \mathcal{A})$ to $(F, \mathcal{F})$.

10.2 Let $\mathbf{R}^2 = \mathbf{R} \times \mathbf{R}$, and let $\mathcal{B}^2$ be the Borel sets of $\mathbf{R}^2$, while $\mathcal{B}$ denotes the Borel sets of $\mathbf{R}$. Show that $\mathcal{B}^2 = \mathcal{B} \otimes \mathcal{B}$.

10.3 Let $\Omega = [0, 1]$, $\mathcal{A}$ be the Borel sets of $[0, 1]$, and let $P(A) = \int 1_A(x)dx$ for $A \in \mathcal{A}$. Let $X(x) = x$. Show that X has the uniform distribution.

10.4 Let $\Omega = \mathbf{R}$ and $\mathcal{A} = \mathcal{B}$. Let P be given by $P(A) = \frac{1}{\sqrt{2\pi}} \int 1_A(x)e^{-x^2/2}dx$. Let $X(x) = x$. Show that X has a normal distribution with parameters $\mu = 0$ and $\sigma^2 = 1$.

10.5 Construct an example to show that $E\{XY\} = E\{X\}E\{Y\}$ does not imply in general that X and Y are independent r.v.'s (we assume X, Y and XY are all in L^1).

10.6 Let X, Y be independent random variables taking values in $\mathbf{N}$ with

$$P(X = i) = P(Y = i) = \frac{1}{2^i} \qquad (i = 1, 2, \ldots).$$

Find the following probabilities:

a) $P(\min(X, Y) \leq i)$ [Ans.: $1 - \frac{1}{4^i}$]
b) $P(X = Y)$ [Ans.: $\frac{1}{3}$]
c) $P(Y > X)$ [Ans.: $\sum_{i \geq 0} \frac{1}{2^i(2^i - 1)}$]
d) $P(X$ divides $Y)$ [Ans.: $\frac{1}{3}$]
e) $P(X \geq kY)$ for a given positive integer k [Ans.: $\frac{1}{2^{1+k}-1}$]

10.7 Let X, Y be independent geometric random variables with parameters λ and μ. Let $Z = \min(X, Y)$. Show Z is geometric and find its parameter. [Ans: $\lambda\mu$.]

10.8 Let $X, Y \in L^2$. Define the *covariance* of X and Y as

$$\text{Cov}(X, Y) = E\{(X - \mu)(Y - \nu)\}$$

where $E\{X\} = \mu$ and $E\{Y\} = \nu$. Show that

$$\text{Cov}(X, Y) = E\{XY\} - \mu\nu$$

and show further that X and Y independent implies $\text{Cov}(X, Y) = 0$.

10.9 Let $X, Y \in L^1$. If X and Y are independent, show that $XY \in L^1$. Give an example to show XY need not be in L^1 in general (i.e., if X and Y are not independent).

10.10 * Let n be a prime number greater than 2; and let X, Y be independent and uniformly distributed on $\{0, 1, \ldots, n-1\}$. (That is, $P(X = i) = P(Y = i) = \frac{1}{n}$, for $i = 0, 1, \ldots, n-1$.) For each r, $0 \leq r \leq n-1$, define $Z_r = X + rY \pmod{n}$.

a) Show that the r.v.'s $\{Z_r : 0 \leq r \leq n-1\}$ are pairwise independent.
b) Is the same result true if n is no longer assumed to be prime? [Ans: No.]

10.11 Let X and Y be independent r.v.'s with distributions $P(X = 1) = P(Y = 1) = \frac{1}{2}$ and $P(X = -1) = P(Y = -1) = \frac{1}{2}$. Let $Z = XY$. Show that X, Y, Z are pairwise independent but that they are not mutually independent.

10.12 Let A_n be a sequence of events. Show that

$$P(A_n \text{ i.o.}) \geq \limsup_{n \to \infty} P(A_n).$$

10.13 A sequence of r.v.'s $X_1, X_2, \ldots$ is said to be *completely convergent* to X if

$$\sum_{n=1}^{\infty} P(|X_n - X| > \varepsilon) < \infty \text{ for each } \varepsilon > 0.$$

Show that if the sequence X_n is independent then complete convergence is equivalent to convergence a.s.

10.14 Let μ, ν be two finite measures on $(E, \mathcal{E})$, $(F, \mathcal{F})$, respectively, i.e. they satisfy all axioms of probability measures except that $\mu(E)$ and $\nu(F)$ are positive reals, but not necessarily equal to 1. Let $\lambda = \mu \otimes \nu$ on $(E \times F, \mathcal{E} \otimes \mathcal{F})$ be defined by $\lambda(A \times B) = \mu(A)\nu(B)$ for Cartesian products $A \times B$ ($A \in \mathcal{E}$, $B \in \mathcal{F}$).

a) Show that λ extends to a finite measure defined on $\mathcal{E} \otimes \mathcal{F}$;
b) Let $f : E \times F \to \mathbf{R}$ be measurable. Prove *Fubini's Theorem:* if f is λ-integrable, then $x \to \int f(x, y)\nu(dy)$ and $y \to \int f(x, y)\mu(dx)$ are respectively $\mathcal{E}$ and $\mathcal{F}$ measurable, and moreover

$$\int f \, d\lambda = \int\!\!\int f(x, y)\mu(dx)\nu(dy) = \int\!\!\int f(x, y)\nu(dy)\mu(dx).$$

(*Hint:* Use Theorem 10.3.)

10.15 * A measure τ is called σ-finite on $(G, \mathcal{G})$ if there exists a sequence of sets $(G_j)_{j \geq 1}$, $G_j \in \mathcal{G}$, such that $\cup_{j=1}^{\infty} G_j = G$ and $\tau(G_j) < \infty$, each j. Show that if μ, ν are assumed to be σ-finite and assuming that $\lambda = \mu \otimes \nu$ exists, then

a) $\lambda = \mu \otimes \nu$ is σ-finite; and

b) (*Fubini's Theorem*): If $f: E \times F \to \mathbf{R}$ is measurable and λ-integrable, then $x \to \int f(x,y)\nu(dy)$ and $y \to \int f(x,y)\mu(dx)$ are respectively $\mathcal{E}$ and $\mathcal{F}$ measurable, and moreover

$$\int f \, d\lambda = \iint f(x,y)\mu(dx)\nu(dy) = \iint f(x,y)\nu(dy)\mu(dx).$$

(*Hint:* Use Exercise 10.14 on sets $E_j \times F_k$, where $\mu(E_j) < \infty$ and $\nu(F_k) < \infty$.)

10.16 * Toss a coin with $P(\text{Heads}) = p$ repeatedly. Let A_k be the event that k or more consecutive heads occurs amongst the tosses numbered $2^k, 2^k + 1, \ldots, 2^{k+1} - 1$. Show that $P(A_k \text{ i.o.}) = 1$ if $p \geq \frac{1}{2}$ and $P(A_k \text{ i.o.}) = 0$ if $p < \frac{1}{2}$.

10.17 Let $X_0, X_1, X_2, \ldots$ be independent random variables with $P(X_n = 1) = P(X_n = -1) = \frac{1}{2}$, all n. Let $Z_n = \Pi_{i=0}^n X_i$. Show that $Z_1, Z_2, Z_3, \ldots$ are independent.

10.18 Let X, Y be independent and suppose $P(X + Y = \alpha) = 1$, where α is a constant. Show that both X and Y are constant random variables.

11 Probability Distributions on R

We have already seen that a probability measure P on $(\mathbf{R}, \mathcal{B})$ (with $\mathcal{B}$ the Borel sets of $\mathbf{R}$) is characterized by its distribution function

$$F(x) = P((-\infty, x]).$$

We now wish to use the tools we have developed to study *Lebesgue measure* on $\mathbf{R}$.

Definition 11.1. Lebesgue measure *is a set function* $m: \mathcal{B} \to [0, \infty]$ *that satisfies*

(i) *(countable additivity) if* $A_1, A_2, A_3, \ldots$ *are pairwise disjoint Borel sets, then*

$$m\left(\cup_{i=1}^{\infty} A_i\right) = \sum_{i=1}^{\infty} m(A_i)$$

(ii) *if* $a, b \in \mathbf{R}$, $a < b$, *then* $m((a, b]) = b - a$.

Theorem 11.1. *Lebesgue measure is unique.*

Proof. Fix $a < b$ in $\mathbf{R}$, and define

$$m_{a,b}(A) = \frac{m(A \cap (a, b])}{b - a}, \quad \text{all } A \in \mathcal{B}.$$

Then $m_{a,b}$ is a probability measure on $(\mathbf{R}, \mathcal{B})$, and the corresponding "distribution" function $F_{a,b}$ is given by

$$F_{a,b}(x) = m_{a,b}((-\infty, x]) = \begin{cases} 0 & \text{if } x < a \\ \dfrac{x - a}{b - a} & \text{if } a \leq x < b \\ 1 & \text{if } b \leq x. \end{cases} \tag{11.1}$$

Therefore $m_{a,b}$ is uniquely determined (since $F_{a,b}$ has a given formula and is thus unique). Moreover since

$$m(A) = \sum_{n \in \mathbf{Z}} m_{n,n+1}(A), \quad \text{any } A \in \mathcal{B}, \tag{11.2}$$

we have that m is uniquely determined as well. $\qquad\square$

Now that we know Lebesgue measure is unique, we need to know it exists!

Theorem 11.2. *Lebesgue measure exists.*

Proof. The function $F_{a,b}$ given in (11.1) clearly exists, and it is nondecreasing, continuous and equals 0 for x small enough, and it equals 1 for x large enough. Therefore the probability measure $m_{a,b}$ also exists by Theorem 7.2. Thus it suffices to define m by (11.2). The verification of countable additivity and that $m((a,b]) = b - a$ are all immediate. □

The theory of integration that we sketched in Chapter 9 remains true for Lebesgue measure: the only difference is that $m(\mathbf{R})$ does not equal 1 but equals $+\infty$: all the results of Chapter 9 remain valid except for the statements that any bounded Borel function is integrable (Theorem 9.1(b)) and that $L^2 \subset L^1$ (Theorem 9.3(b)), which are now false!

If f is a Borel measurable function which is integrable for Lebesgue measure, then its integral is written $\int f(x)dx$. Recall that f is integrable if $\int f^+(x)dx < \infty$ and $\int f^-(x)dx < \infty$, where $f = f^+ - f^-$. The Lebesgue integral exists more generally than does the Riemann integral, but when they both exist then they are equal.

Definition 11.2. *The* density *of a probability measure P on $(\mathbf{R}, \mathcal{B})$ is a positive Borel measurable function f that verifies for all $x \in \mathbf{R}$:*

$$P((-\infty, x]) = \int_{-\infty}^{x} f(y)dy = \int f(y)1_{(-\infty, x]}(y)dy. \qquad (11.3)$$

If $P = P^X$, the distribution measure of a r.v. X, then we say f is the density of X.

Warning: As already stated in Chapter 7, not all probability measures on $(\mathbf{R}, \mathcal{B})$ have densities. Indeed, (11.3) implies that F is *continuous,* and not all F are continuous. Actually (11.3) is much stronger than continuity, and there are even continuous distribution functions F whose corresponding probabilities do not have densities.

Theorem 11.3. *A positive Borel measurable function f on $\mathbf{R}$ is the density of a probability measure on $(\mathbf{R}, \mathcal{B})$ if and only if it satisfies $\int f(x)dx = 1$. In this case it entirely determines the probability measure, and any other positive Borel measurable function f' such that $m(f \neq f') = 0$ is also a density for the same probability measure.*

Conversely a probability measure on $(\mathbf{R}, \mathcal{B})$ determines its density (when it exists) up to a set of Lebesgue measure zero (i.e., if f and f' are two densities for this probability, then $m(f \neq f') = 0$).

Proof. Let f be a density for a probability measure P. By (11.3) we have

$$\int_{-\infty}^{x} f(y)dy = P((-\infty, x]).$$

Let x increase to ∞, and we see that

$$\int f(y)dy = \int_{-\infty}^{\infty} f(y)dy = \lim_{x \to \infty} \int_{-\infty}^{x} f(y)dy = 1.$$

Thus $\int f(x)dx = 1$.

For the sufficient condition one could give a very short proof based upon the distribution function. Nevertheless we give a longer but more direct proof which readily extends to probabilities on $\mathbf{R}^n$. Let f be a positive Borel function with $\int f(x)dx = 1$. For every Borel set A we put

$$P(A) = \int_A f(y)dy = \int f(y)1_A(y)dy. \tag{11.4}$$

This defines a function $P : \mathcal{B} \to \mathbf{R}_+$ which clearly has $P(\mathbf{R}) = 1$. Further if $A_1, A_2, \ldots, A_m, \ldots$ are all pairwise disjoint, then

$$P\left(\cup_{i=1}^{\infty} A_i\right) = \int f(x)1_{\{\cup_{i=1}^{\infty} A_i\}}(x)dx$$

$$= \int \left(\sum_{i=1}^{\infty} f(x)1_{A_i}(x)\right) dx$$

since the A_i are pairwise disjoint;

$$= \sum_{i=1}^{\infty} \int f(x)1_{A_i}(x)dx = \sum_{i=1}^{\infty} P(A_i)$$

by using Theorem 9.2. Therefore we have countable additivity and P is a true probability measure on $(\mathbf{R}, \mathcal{B})$. Taking $A = (-\infty, x]$ in (11.4) yields

$$P((-\infty, x]) = \int_{-\infty}^{x} f(y)dy,$$

that is P admits the density f.

We now show that P determines f up to a set of Lebesgue measure zero. Suppose f' is another density for P. Then f' will also satisfy (11.4) (to see this, define P' by (11.4) with f' and observe that both P and P' have the same distribution function, implying that $P = P'$). Therefore, if we choose $\varepsilon > 0$ and set $A = \{x : f(x) + \varepsilon \le f'(x)\}$ and if $m(A) > 0$, then

$$P(A) + \varepsilon m(A) = \int (f(x) + \varepsilon)1_A(x)dx \le \int f'(x)1_A(x)dx = P(A),$$

a contradiction. We conclude $m(\{f + \varepsilon \le f'\}) = 0$. Since $\{f + \varepsilon \le f'\}$ increases to $\{f < f'\}$ as ε decreases to 0, we obtain that $m(\{f' < f\}) = 0$. Analogously, $m(\{f' > f\}) = 0$, hence $f' = f$ almost everywhere (dm). ["Almost everywhere" means except on a set of measure zero; for probability measures we say "almost surely" instead of "almost everywhere".] $\quad\square$

Remark 11.1. Since the density f and the distribution function F satisfy $F(x) = \int_{-\infty}^{x} f(y)dy$, one is tempted to conclude that F is differentiable, with derivative equal to f. This is true at each point x where f is continuous. One can show – and this is a difficult result due to Lebesgue – that F is differentiable dm-almost everywhere regardless of the nature of f. But this result is an almost everywhere result, and it is *not* true in general for *all* x. However in most "concrete" examples, when the density exists it turns out that F is piecewise differentiable: in this case one may take $f = F'$ (the derivative of F) wherever it exists, and $f = 0$ elsewhere.

Corollary 11.1 (Expectation Rule). *Let X be an **R**-valued r.v. with density f. Let g be a Borel measurable function. Then g is integrable (resp. admits an integral) with respect to P^X, the distribution measure of X, if and only if the product fg is integrable (resp. admits an integral) with respect to Lebesgue measure, and in this case we have*

$$E\{g(X)\} = \int g(x)P^X(dx) = \int g(x)f(x)dx. \tag{11.5}$$

Proof. The equality (11.5) holds for indicator functions by Theorem 11.3, because it reduces to (11.4). Therefore (11.5) holds for simple functions by linearity. For g nonnegative, let g_n be simple functions increasing to g. Then (11.5) holds by the monotone convergence theorem. For general g, let $g = g^+ - g^-$, and the result follows by taking differences.   □

We presented examples of densities in Chapter 7. Note that all the examples were continuous or piecewise continuous, while here we seem concerned with Borel measurable densities. Most practical examples of r.v.'s in Statistics turn out to have relatively smooth densities, but when we perform simple operations on random variables with nice densities (such as taking a conditional expectation), we quickly have need for a much more general theory that includes Borel measurable densities.

Let X be a r.v. with density f. Suppose $Y = g(X)$ for some g. Can we express the density of Y (if it exists) in terms of f? We can indeed in some "good" cases. We begin with a simple result:

Theorem 11.4. *Let X have density f_X and let g be a Borel measurable function. Let $Y = g(X)$. Then*

$$F_Y(y) = P(Y \le y) = \int_{A_y} f_X(u)du$$

where $A_y = \{u : g(u) \le y\}$.

Note that if F_Y is differentiable, we can use Theorem 11.4 to find the density.

Example: Let X be uniform on $[0, 1]$ and let $Y = -\frac{1}{\lambda}\log(X)$, where $\lambda > 0$. Then

$$F_Y(y) = P\left(-\frac{1}{\lambda}\log(X) \le y\right)$$
$$= P(\log(X) \ge -\lambda y)$$
$$= P(X \ge \exp(-\lambda y))$$
$$= \begin{cases} 1 - e^{-\lambda y} & \text{for } y \ge 0 \\ 0 & \text{otherwise.} \end{cases}$$

Therefore (cf. Remark 11.1):

$$f_Y(y) = \frac{d}{dy}F_Y(y) = \begin{cases} \lambda e^{-\lambda y} & \text{if } y > 0 \\ 0 & \text{if } y \le 0 \end{cases}$$

and we see that Y is exponential with parameter λ.

Caution: The preceding example is deceptively simple because g was injective, or one to one. The general result is given below:

Corollary 11.2. *Let X have a continuous density f_X. Let $g: \mathbf{R} \to \mathbf{R}$ be continuously differentiable with a non-vanishing derivative (hence g is strictly monotone). Let $h(y) = g^{-1}(y)$ be the inverse function (also continuously differentiable). Then $Y = g(X)$ has the density*

$$f_Y(y) = f_X(h(y))|h'(y)|.$$

Proof. Suppose g is increasing. Let $F_Y(y) = P(Y \le y)$. Then

$$F_Y(y) = P(g(X) \le y) = P(h(g(X)) \le h(y)),$$

since h is monotone increasing because g is. Then the above gives

$$= P(X \le h(y)) = F_X(h(y)) = \int_{-\infty}^{h(y)} f(x)dx.$$

It is a standard result from calculus (see, e.g., [18, p.259]) that if a function g is injective (one–to–one), differentiable, and such that its derivative is never zero, then $h = g^{-1}$ is also differentiable and $h'(x) = \frac{1}{f'(h(x))}$. Therefore $F_Y(y)$ is differentiable and

$$\frac{d}{dy}F_Y(y) = f(h(y))h'(y) = f(h(y))|h'(y)|.$$

If g is decreasing the same argument yields

$$\frac{d}{dy}F_Y(y) = f(h(y)))(-h'(y)) = f(h(y))|h'(y)|.$$

□

Corollary 11.3. *Let X have a continuous density f_X. Let $g: \mathbf{R} \to \mathbf{R}$ be piecewise strictly monotone and continuously differentiable: that is, there exist intervals $I_1, I_2, \ldots, I_n$ which partition $\mathbf{R}$ such that g is strictly monotone and continuously differentiable on the interior of each I_i. For each i, $g : I_i \to \mathbf{R}$ is invertible on $g(I_i)$ and let h_i be the inverse function. Let $Y = g(X)$ and let $\wedge = \{y \colon y = g(x), x \in \mathbf{R}\}$, the range of g. Then the density f_Y of Y exists and is given by*

$$f_Y(y) = \sum_{i=1}^{n} f_X(h_i(y)) |h_i'(y)| 1_{g(I_i)}(y).$$

Remark: The proof, similar to the proof of the previous corollary, is left to the reader. Our method uses the continuity of f_X, but the result holds when f_X is simply measurable.

Example: Let X be normal with parameters $\mu = 0$; $\sigma^2 = 1$. Let $Y = X^2$. Then in this case $g(x) = x^2$, which is neither monotone nor injective. Take $I_1 = [0, \infty)$ and $I_2 = (-\infty, 0)$. Then g is injective and strictly monotone on I_1 and I_2, and $I_1 \cup I_2 = \mathbf{R}$. $g(I_1) = [0, \infty)$ and $g(I_2) = (0, \infty)$. Then $h_1 : [0, \infty) \to \mathbf{R}$ by $h_1(y) = \sqrt{y}$ and $h_2 : [0, \infty) \to \mathbf{R}$ by $h_2(y) = -\sqrt{y}$.

$$|h_i'(y)| = \left| \frac{1}{2\sqrt{y}} \right| = \frac{1}{2\sqrt{y}}, \text{ for } i = 1, 2.$$

Therefore by Corollary 11.3,

$$f_Y(y) = \frac{1}{\sqrt{2\pi}} e^{-y/2} \frac{1}{2\sqrt{y}} + \frac{1}{\sqrt{2\pi}} e^{-y/2} \frac{1}{2\sqrt{y}} \qquad (y > 0)$$

$$= \frac{1}{\sqrt{2\pi}} \frac{1}{\sqrt{y}} e^{-y/2} 1_{(0,\infty)}(y).$$

The random variable Y is called a χ^2 random variable with one degree of freedom. (This is pronounced "chi square".)

The preceding example is sufficiently simple that it can also be derived "by hand", without using Corollary 11.3. Indeed,

$$F_Y(y) = P(Y \le y) = P(X^2 \le y)$$
$$= P(-\sqrt{y} \le X \le \sqrt{y})$$
$$= F_X(\sqrt{y}) - F_X(-\sqrt{y});$$

and

$$F_X(\sqrt{y}) = \int_{-\infty}^{\sqrt{y}} \frac{1}{\sqrt{2\pi}} e^{-x^2/2} dx.$$

Thus differentiating yields

$$\frac{d}{dy} F_X(\sqrt{y}) = \frac{1}{\sqrt{2\pi}} e^{-y/2} \frac{1}{2\sqrt{y}} 1_{(y>0)}.$$

Similarly,

$$-F_X\left(-\sqrt{y}\right) = -\int_{-\infty}^{-\sqrt{y}} \frac{1}{\sqrt{2\pi}} e^{-x^2/2} dx,$$

whence

$$\frac{d}{dy}\left(-F\left(-\sqrt{y}\right)\right) = -\frac{1}{\sqrt{2\pi}} e^{-y/2} \frac{-1}{2\sqrt{y}} 1_{(y>0)}$$

$$= \frac{1}{\sqrt{2\pi}} e^{-y/2} \frac{1}{2\sqrt{y}} 1_{(y>0)},$$

and adding yields the same result as we obtained using the Corollary.

Remark: The *chi square distribution* plays an important role in Statistics. Let p be an integer. Then a random variable X with density

$$f(x) = \frac{1}{\Gamma(p/2)2^{p/2}} x^{p/2-1} e^{-\frac{x}{2}}, \qquad 0 < x < \infty$$

is called a *chi squared density with p degrees of freedom*. This is usually denoted χ_p^2. Note that it is a special case of a Gamma distribution: indeed X is also Gamma $(\frac{p}{2}, \frac{1}{2})$. We have just seen in the example that if X is χ_1^2, then X equals Z^2 in distribution, where Z is $N(0,1)$. We will see in Chapter 15 (Example 6) that if X is χ_p^2 then $X = \sum_{i=1}^p Z_i^2$ in distribution, where Z_i are i.i.d. $N(0,1)$, $1 \le i \le p$. Such a distribution arises naturally in Statistics when one tries to estimate the (unknown) variance of a normally distributed population. (See in this regard Exercise 15.13).

Let us also note that a χ_2^2 is simply an exponential random variable with parameter $\lambda = \frac{1}{2}$.

Exercises for Chapter 11

11.1 Use the density for a chi square r.v. to show that $\Gamma(\frac{1}{2}) = \sqrt{\pi}$.

11.2 Let X be uniformly distributed on $[-1, 1]$. Find the density of $Y = X^k$ for positive integers k. [Ans: for k odd, $f_Y(y) = \frac{1}{2k}y^{\frac{1}{k}-1}1_{[-1,1]}(y)$; for k even, $f_Y(y) = \frac{1}{k}y^{\frac{1}{k}-1}1_{[0,1]}(y)$.]

11.3 Let X have distribution function F. What is the distribution function of $Y = |X|$? When X admits a continuous density f_X, show that Y also admits a density f_Y, and express f_Y in terms of f_X.

11.4 Let X be Cauchy with parameters α, 1. Let $Y = \frac{a}{X}$ with $a \neq 0$. Show Y is also a Cauchy r.v. and find its parameters. [Ans: $\frac{a\alpha}{1+\alpha^2}, \sqrt{\frac{|a|}{1+\alpha^2}}$].

11.5 Let X have a density f_X, and let $Y = \frac{a}{X}$ with $a \neq 0$. Find the density of Y in terms of f_X. [Ans: $f_Y(y) = \frac{|a|}{y^2}f_X(\frac{a}{y})$.]

11.6 Let X be uniform on $(-\pi, \pi)$, and let $Y = \sin(X + \theta)$. Show that the density for Y is $f_Y(y) = \frac{2}{2\pi\sqrt{1-y^2}}1_{[-1,1]}(y)$.

11.7 * Let X have a density and let $Y = a\sin(X + \theta)$, $a > 0$. Show that:

$$f_Y(y) = \frac{1}{\sqrt{a^2 - y^2}}\sum_{i=-\infty}^{\infty}(f_X(h_i(y)) + f_X(k_i(y))1_{[-a,a]}(y)$$

for appropriate functions h_i and k_i.

11.8 * Let X be uniform on $(-\pi, \pi)$ and let $Y = a\tan(X)$, $a > 0$. Find $f_Y(y)$. [Ans: $f_Y(y) = \frac{a/\pi}{a^2+y^2}$.]

11.9 * Let X have a density, and let

$$Y = ce^{-\alpha X}1_{\{X>0\}}, \qquad (\alpha > 0, c > 0).$$

Find $f_Y(y)$ in terms of f_X. [Ans: $f_Y(y) = \frac{f_X(-\frac{1}{\alpha}\ln(\frac{y}{c}))}{\alpha y}1_{(0,c)}(y)$.]

11.10 A *density f is called symmetric* if $f(-x) = f(x)$, for all x. (That is, f is an even function.) A *random variable X is symmetric* if X and $-X$ both have the same distribution. Suppose X has a density f. Show that X is symmetric if and only if it has a density f which is symmetric. In this case, does it admit also a nonsymmetric density? [Ans.: Yes, just modify f on a non-empty set of Lebesgue measure zero in $\mathbf{R}_+$]. [Note: Examples of symmetric densities are the uniform on $(-a, a)$; the normal with parameters $(0, \sigma^2)$; double exponential with parameters $(0, \beta)$; the Cauchy with parameters $(0, \beta)$.]

11.11 Let X be positive with a density f. Let $Y = \frac{1}{X+1}$ and find the density for Y.

11.12 Let X be normal with parameters (μ, σ^2). Show that $Y = e^X$ has a lognormal distribution.

11.13 Let X be a r.v. with distribution function F that is continuous. Show that $Y = F(X)$ is uniform.

11.14 Let F be a distribution function that is continuous and is such that the inverse function F^{-1} exists. Let U be uniform on $(0, 1)$. Show that $X = F^{-1}(U)$ has distribution function F.

11.15 * Let F be a continuous distribution function and let U be uniform on $(0, 1)$. Define $G(u) = \inf\{x : F(x) \geq u\}$. Show that $G(U)$ has distribution function F.

11.16 Let $Y = -\frac{1}{\lambda}\ln(U)$, where U is uniform on $(0, 1)$. Show that Y is exponential with parameter λ by inverting the distribution function of the exponential. (*Hint:* If U is uniform on $(0, 1)$ then so also is $1 - U$.) This gives a method to simulate exponential random variables.

12 Probability Distributions on $\mathbf{R}^n$

In Chapter 11 we considered the simple case of distributions on $(\mathbf{R}, \mathcal{B})$. The case of distributions on $(\mathbf{R}^n, \mathcal{B}^n)$ for $n = 2, 3, \ldots$ is both analogous and more complicated. [$\mathcal{B}^n$ denotes the Borel sets of $\mathbf{R}^n$.]

First let us note that by essentially the same proof as used in Theorem 2.1, we have that $\mathcal{B}^n$ is generated by "quadrants" of the form

$$\prod_{i=1}^{n} (-\infty, a_i]; \qquad a_i \in \mathbf{Q};$$

note that $\mathcal{B} \otimes \mathcal{B} \otimes \ldots \otimes \mathcal{B} = \mathcal{B}^n$; that is, $\mathcal{B}^n$ is also the smallest σ-algebra generated by the n-fold Cartesian product of $\mathcal{B}$, the Borel sets on $\mathbf{R}$.

The *n-dimensional distribution function* of a probability measure on $(\mathbf{R}^n, \mathcal{B}^n)$ is defined to be:

$$F(x_1, \ldots, x_n) = P\left(\prod_{i=1}^{n}(-\infty, x_i]\right).$$

It is more subtle to try to characterize P by using F for $n \geq 2$ than it is for $n = 1$, and consequently distribution functions are rarely used for $n \geq 2$.

We have also seen that the density of a probability measure on $\mathbf{R}$, when it exists, is a very convenient tool. Contrary to distribution functions, this notion of a density function extends easily and is exactly as convenient on $\mathbf{R}^n$ as it is on $\mathbf{R}$ (but, as is the case for $n = 1$, it does not always exist).

Definition 12.1. *The* Lebesgue measure m_n *on* $(\mathbf{R}^n, \mathcal{B}^n)$ *is defined on Cartesian product sets* $A_1 \times A_2 \times \ldots \times A_n$ *by*

$$m_n\left(\prod_{i=1}^{n} A_i\right) = \prod_{i=1}^{n} m(A_i), \qquad \text{all } A_i \in \mathcal{B}, \tag{12.1}$$

where m is the one dimensional Lebesgue measure defined on $(\mathbf{R}, \mathcal{B})$. As in Theorem 10.3, one can extend the measure defined in (12.1) for Cartesian product sets uniquely to a measure m_n on $(\mathbf{R}^n, \mathcal{B}^n)$, and m_n will still have countable additivity. This measure m_n is Lebesgue measure, and it is characterized also by the following seemingly weaker condition than (12.1):

$$m_n\left(\prod_{i=1}^n (a_i, b_i]\right) = \prod_{i=1}^n (b_i - a_i), \qquad all \quad -\infty < a_i < b_i < \infty.$$

If $A \in \mathcal{B}^n$, one can view $m_n(A)$ as the "volume" of the set A, a property which is apparent for "rectangles" of the form (12.1).

We write

$$\int f(x)dx = \int f(x_1, \ldots, x_n)dx_1 dx_2 \ldots dx_n$$

to denote the integral of f with respect to m_n, and also $\int_A f(x)dx$ for the integral of the product $f1_A$ when $A \in \mathcal{B}^n$, as in the one-dimensional case.

Definition 12.2. *A probability measure P on $(\mathbf{R}^n, \mathcal{B}^n)$ has a density f if f is a nonnegative Borel measurable function on $\mathbf{R}^n$ verifying*

$$P(A) = \int_A f(x)dx = \int f(x)1_A(x)dx$$
$$= \int f(x_1, \ldots, x_n)1_A(x_1, \ldots, x_n)dx_1 \ldots dx_n,$$

for all $A \in \mathcal{B}^n$.

Once more, we warn the reader that not all probabilities on $(\mathbf{R}^n, \mathcal{B}^n)$ have densities!

The next theorem is the exact analogue of Theorem 11.3, and the proof is similar and is not repeated here:

Theorem 12.1. *A positive Borel measurable function f on $\mathbf{R}^n$ is the density of a probability measure on $(\mathbf{R}^n, \mathcal{B}^n)$ if and only if it satisfies $\int f(x)dx = 1$. In this case it entirely determines the probability measure, and any other positive Borel measurable function f' such that $m_n(f \neq f') = 0$ is also a density for the same probability measure.*

Conversely a probability measure on $(\mathbf{R}^n, \mathcal{B}^n)$ determines its density (when it exists) up to a set of Lebesgue measure zero (i.e., if f and f' are two densities for this probability, then $m_n(f \neq f') = 0$; we also write: $f = f'$ m_n-a.e.).

For simplicity, we now let $n = 2$. That is, we restrict our discussion to random variables taking their values in $\mathbf{R}^2$; Theorem 12.2 below generalizes easily to $\mathbf{R}^n$, $n = 3, 4, \ldots$.

Let X be an $\mathbf{R}^2$-valued r.v. with components Y and Z; that is, $X = (Y, Z)$.

Theorem 12.2. *Assume that $X = (Y, Z)$ has a density f on $\mathbf{R}^2$. Then:*

a) *Both Y and Z have densities on $(\mathbf{R}, \mathcal{B})$ given by:*

$$f_Y(y) = \int_{-\infty}^{\infty} f(y, z)dz; \qquad f_Z(z) = \int_{-\infty}^{\infty} f(y, z)dy. \qquad (12.2)$$

b) *Y and Z are independent if and only if*

$$f(y, z) = f_Y(y)f_Z(z) \qquad (dm_2 \text{ a.e.}).$$

c) *The formula below defines another density on $\mathbf{R}$ at every point $y \in \mathbf{R}$ such that $f_Y(y) \neq 0$:*

$$f_{Y=y}(z) = \frac{f(y, z)}{f_Y(y)}.$$

Before we prove Theorem 12.2 we digress a bit to explain part (c) above: the densities $f_Y(y)$ and $f_Z(z)$ are called the *marginal densities of f*. Note that one cannot in general recover what f is from knowledge of the marginals alone. (The exception is when Y and Z are independent.) Thus the *"joint density"* f of $X = (Y, Z)$ in general contains more information than do the two marginals.

The function $f_{Y=y}(z)$ is called the *"conditional density of Z given $Y = y$"*. This cannot be interpreted literally, because here $P(Y = y) = 0$ for each y, and conditional probabilities of the type $P(A \mid Y = y)$ have no meaning when $P(Y = y) = 0$. Nevertheless the terminology has a heuristic justification as follows. Let Δy and Δz denote very small changes in y and z. Then

$$f(y, z)\Delta y\Delta z \approx P(y \leq Y \leq y + \Delta y; z \leq Z \leq z + \Delta z)$$

and

$$f_Y(y)\Delta y \approx P(y \leq Y \leq y + \Delta y);$$

in this case $P(y \leq Y \leq y + \Delta y)$ can be assumed to be strictly positive, and then by division:

$$\begin{aligned}
f_{Y=y}(z)\Delta z &\approx \frac{P(y \leq Y \leq y + \Delta y; z \leq Z \leq z + \Delta z)}{P(y \leq Y \leq y + \Delta y)} \\
&\approx P(z \leq Z \leq z + \Delta z \mid Y \approx y).
\end{aligned}$$

Proof of Theorem 12.2: a) For each Borel set $A \in \mathcal{B}$, we have

$$\begin{aligned}
P(Y \in A) = P(X \in A \times \mathbf{R}) &= \iint_{A \times \mathbf{R}} f(y, z)dy\, dz \\
&= \int_A dy \int_{-\infty}^{\infty} f(y, z)dz \\
&= \int_A dy\, f_Y(y),
\end{aligned}$$

and since this holds for all $A \in \mathcal{B}$, and since densities on $\mathbf{R}$ are characterized by (11.4), f_Y as defined in (12.1) is a density of Y. The proof for f_Z is the same.

b) Suppose $f(y,z) = f_Y(y)f_Z(z)$. Then

$$
\begin{aligned}
P(Y \in A, Z \in B) &= \iint 1_{A \times B}(y,z)f(y,z)dy\,dz \\
&= \iint 1_A(y)1_B(z)f(y,z)dy\,dz \\
&= \iint 1_A(y)1_B(z)f_Y(y)f_Z(z)dy\,dz \\
&= \int 1_A(y)f_Y(y)dy \int 1_B(z)f_Z(z)dz \\
&= P(Y \in A)P(Z \in B),
\end{aligned}
$$

and since A, B are arbitrary Borel sets, we have Y and Z are independent. Now suppose Y and Z are independent. Let

$$
\mathcal{H} = \left\{ C \in \mathcal{B}^2 : \iint_C f(y,z)dy\,dz = \iint_C f_Y(y)f_Z(z)dy\,dz \right\}.
$$

Then since Y, Z are independent, if $C = A \times B$ with $A \in \mathcal{B}$ and $B \in \mathcal{B}$, then

$$
P((Y,Z) \in C) = \iint_C f(y,z)dy\,dz
$$

while

$$
\begin{aligned}
P((Y,Z) \in C) &= P(Y \in A, Z \in B) \\
&= P(Y \in A)P(Z \in B) \\
&= \int_A f_Y(y)dy \int_B f_z(z)dz \\
&= \iint_{A \times B} f_Y(y)f_Z(z)dy\,dz
\end{aligned}
$$

by the Tonelli-Fubini Theorem (Theorem 10.3). Therefore $\mathcal{H}$ contains the class of all products $C = A \times B$ where $A, B \in \mathcal{B}$, while this latter class is closed under finite intersections and generates the σ-algebra $\mathcal{B}^2$. Since further $\mathcal{H}$ is closed under increasing limits and differences, we deduce from Theorem 6.2 that $\mathcal{H} = \mathcal{B}^2$. Therefore

$$
P(X \in C) = \int_C f(y,z)dy\,dz = \int_C f_Y(y)f_Z(z)dy\,dz
$$

for all Borel sets $C \in \mathcal{B}^2$. Then the uniqueness of the density (Theorem 12.1) gives

$$
f(y,z) = f_Y(y)f_Z(z), \qquad \text{a.e. } dm_2.
$$

c) We have

$$\int f_{Y=y}(z)dz = \int_{-\infty}^{\infty} \frac{f(y,z)}{f_Y(y)}dz$$
$$= \frac{1}{f_Y(y)} \int_{-\infty}^{\infty} f(y,z)dz = \frac{1}{f_Y(y)} f_Y(y) = 1.$$

Since $f_{Y=y}(z)$ is positive, Borel measurable, and integrates to 1, it is a density. $\qquad\square$

Definition 12.3. *Let X, Y be two real valued random variables, each with finite variance. The* covariance *of X, Y is defined to be*

$$\mathrm{Cov}(X,Y) = E\{(X - E\{X\})(Y - E\{Y\})\} = E\{XY\} - E\{X\}E\{Y\}.$$

Note that $E\{XY\}$ exists: since X and Y have finite variances they are both in L^2, and Theorem 9.3a then gives that $XY \in L^1$. We remark that

$$\mathrm{Cov}(X,X) = \mathrm{Var}(X) = \sigma^2(X).$$

Theorem 12.3. *If X and Y are independent, then $\mathrm{Cov}(X,Y) = 0$.*

Proof. X and Y independent implies that $E\{XY\} = E\{X\}E\{Y\}$, and the result follows. $\qquad\square$

Warning: The converse to Theorem 12.3 is false in general: if one has $\mathrm{Cov}(X,Y) = 0$ it is not true in general that X and Y are independent.

Definition 12.4. *Let X and Y be two r.v.'s, both with finite variance. The* correlation coefficient *of X and Y is the number*

$$\rho = \frac{\mathrm{Cov}(X,Y)}{\sigma(X)\sigma(Y)}.$$

($\sigma(X) = \sqrt{\sigma^2(X)}$ and $\sigma(Y) = \sqrt{\sigma^2(Y)}$.)

Note that by the Cauchy-Schwarz inequality (Theorem 9.3a) we have always that $-1 \le \rho \le 1$, and if X and Y are independent then $\rho = 0$ by Theorem 12.3.

Definition 12.5. *Let $X = (X_1, \ldots, X_n)$ be an $\mathbf{R}^n$-valued random variable. The* covariance matrix *of X is the $n \times n$ matrix whose general term is*

$$c_{ij} = \mathrm{Cov}(X_i, X_j).$$

Theorem 12.4. *A covariance matrix is positive semidefinite; that is, it is symmetric ($c_{ij} = c_{ji}$, all i, j) and also $\sum a_i a_j c_{ij} \ge 0$, for all $(a_1, \ldots, a_n) \in \mathbf{R}^n$.*

Proof. The symmetry is clear, since $\mathrm{Cov}(X_i, X_j) = \mathrm{Cov}(X_j, X_i)$ trivially. A simple calculation shows that

$$\sum a_i a_j c_{ij} = \mathrm{Var}\left(\sum_{i=1}^n a_i X_i\right),$$

and since variances are always nonnegative, we are done.   □

Theorem 12.5. *Let X be an $\mathbf{R}^n$-valued r.v. with covariance matrix C. Let A be an $m \times n$ matrix and set $Y = AX$. Then Y is an $\mathbf{R}^m$-valued r.v. and its covariance matrix is $C' = ACA^*$, where A^* denotes A transpose.*

Proof. The proof is a simple calculation.   □

We now turn our attention to functions of $\mathbf{R}^n$-valued random variables. We address the following problem: let $g \colon \mathbf{R}^n \to \mathbf{R}^n$ be Borel. Given $X = (X_1, \ldots, X_n)$ with density f, what is the density of $Y = g(X)$ in terms of f, and to begin with, does it exist at all? We will need the following theorem from advanced calculus (see for example [22, p.83]).

Let us recall first that if g is a differentiable function from an open set G in $\mathbf{R}^n$ into $\mathbf{R}^n$, its *Jacobian matrix* $J_g(x)$ at point $x \in G$ is $J_g(x) = \frac{\partial g}{\partial x}(x)$ (that is, $J_g(x)_{ij} = \frac{\partial g_i}{\partial x_j}(x)$, where $g = (g_1, g_2, \ldots, g_n)$). The Jacobian of g at point x is the determinant of the matrix $J_g(x)$. If this Jacobian is not zero, then g is invertible on a neighborhood of x, and the Jabobian of the inverse g^{-1} at point $y = g(x)$ is the inverse of the Jacobian of g at x.

Theorem 12.6 (Jacobi's Transformation Formula). *Let G be an open set in $\mathbf{R}^n$ and let $g \colon G \to \mathbf{R}^n$ be continuously differentiable.*[1] *Suppose g is injective (one to one) on G and its Jacobian never vanishes. Then for f measurable and such that the product $f1_{g(G)}$ is positive or integrable with respect to Lebesgue measure,*

$$\int_{g(G)} f(y)dy = \int_G f(g(x))|\det(J_g(x))|dx$$

where by $g(G)$ we mean:

$$g(G) = \{y \in \mathbf{R}^n \colon \text{ there exists } x \in G \text{ with } g(x) = y\}.$$

The next theorem is simply an application of Theorem 12.6 to the density functions of random variables.

Theorem 12.7. *Let $X = (X_1, \ldots, X_n)$ have joint density f. Let $g \colon \mathbf{R}^n \to \mathbf{R}^n$ be continuously differentiable and injective, with non-vanishing Jacobian. Then $Y = g(X)$ has density*

---

[1] A function g is *continuously differentiable* if it is differentiable and also its derivative is continuous.

$$f_Y(y) = \begin{cases} f_X(g^{-1}(y))|\det J_{g^{-1}}(y)| & \text{if } y \text{ is in the range of } g \\ 0 & \text{otherwise.} \end{cases}$$

Proof. We denote by G the range of g, that is $G = \{y \in \mathbf{R}^n$: there exists $x \in \mathbf{R}^n$ with $y = g(x)\}$. The properties of g imply that G is an open set and that the inverse function g^{-1} is well defined on G and continuously differentiable with non-vanishing Jacobian. Let $B \in \mathcal{B}^n$, and $A = g^{-1}(B)$. We have

$$P(X \in A) = \int_A f_X(x)dx$$

$$= \int_{g^{-1}(B)} f_X(x)dx$$

$$= \int_B f_X(g^{-1}(x))|\det J_{g^{-1}}(x)|dx,$$

by Theorem 12.6 applied with g^{-1}. But we also have $P(Y \in B) = P(X \in A)$, hence

$$P(X \in A) = \int_B f_Y(y)dy.$$

Since $B \in \mathcal{B}^n$ is arbitrary we conclude

$$f_X(g^{-1}(x))|\det J_{g^{-1}}(x))| = f_Y(x),$$

a.e., whence the result.   □

In analogy to Corollary 11.3 of Chapter 11, we can also treat a case where g is not injective but nevertheless smooth.

Corollary 12.1. *Let $S \in \mathcal{B}^n$ be partitioned into disjoint subsets $S_0, S_1, \ldots, S_m$ such that $\cup_{i=0}^m S_i = S$, and such that $m_n(S_0) = 0$ and that for each $i = 1, \ldots, m$, $g: S_i \to \mathbf{R}^n$ is injective (one to one) and continuously differentiable with non-vanishing Jacobian. Let $Y = g(X)$, where X is an $\mathbf{R}^n$-valued r.v. with values in S and with density f_X. Then Y has a density given by*

$$f_Y(y) = \sum_{i=1}^m f_X(g_i^{-1}(y))|\det J_{g_i^{-1}}(y)|$$

where g_i^{-1} denotes the inverse map $g_i^{-1}: g(S_i) \to S_i$ and $J_{g_i^{-1}}$ is its corresponding Jacobian matrix.

Examples:

1. Let X, Y be independent normal r.v.'s, each with parameters $\mu = 0$, $\sigma^2 = 1$. Let us calculate the joint distribution of $(U, V) = (X+Y, X-Y)$. Here

$$g(x, y) = (x + y, x - y) = (u, v),$$

and

$$g^{-1}(u, v) = \left(\frac{u + v}{2}, \frac{u - v}{2} \right).$$

The Jacobian in this simple case does not depend on (u, v) (that is, it is constant), and is

$$J_{g^{-1}}(u, v) = \begin{pmatrix} \dfrac{1}{2} & \dfrac{1}{2} \\[2mm] \dfrac{1}{2} & -\dfrac{1}{2} \end{pmatrix},$$

and

$$\det J_{g^{-1}} = \frac{1}{2} \left(-\frac{1}{2} \right) - \left(\frac{1}{2} \right) \left(\frac{1}{2} \right) = -\frac{1}{2}.$$

Therefore

$$f_{(U,V)}(u, v) = f_{(X,Y)} \left(\frac{u + v}{2}, \frac{u - v}{2} \right) |\det J|$$

$$= f_X \left(\frac{u + v}{2} \right) f_Y \left(\frac{u - v}{2} \right) |J|$$

$$= \frac{1}{\sqrt{2\pi}} e^{-\frac{1}{2} \left(\frac{u+v}{2} \right)^2} \frac{1}{\sqrt{2\pi}} e^{-\frac{1}{2} \left(\frac{u-v}{2} \right)^2} \cdot \frac{1}{2}$$

$$= \frac{1}{\sqrt{4\pi}} e^{-\frac{u^2}{4}} \frac{1}{\sqrt{4\pi}} e^{-\frac{v^2}{4}}$$

for $-\infty < u, v < \infty$. We conclude that U, V are also independent normals, each with parameters $\mu = 0$ and $\sigma^2 = 2$.

2. Let (X, Y) have joint density f. We want to find the density of $Z = XY$. In this case $h(x, y) = xy$ maps $\mathbf{R}^2$ to $\mathbf{R}^1$, and it appears we cannot use Theorem 12.7 We can however by using a simple trick. Define

$$g(x, y) = (xy, x).$$

We can write $S_0 = \{(x, y) : x = 0, y \in \mathbf{R}\}$ and $S_1 = \mathbf{R}^2 \backslash S_0$. Then $m_2(S_0) = 0$ and g is injective from S_1 to $\mathbf{R}^2$ and $g^{-1}(u, v) = (v, \frac{u}{v})$. The Jacobian

$$J_{g-1}(u, v) = \begin{pmatrix} 0 & \dfrac{1}{v} \\[2mm] 1 & -\dfrac{u}{v^2} \end{pmatrix},$$

and $\det(J_{g-1}) = -\frac{1}{v}$. Therefore Corollary 12.1 gives

$$f_{(U,V)} = \begin{cases} f_{(X,Y)} \left(v, \dfrac{u}{v} \right) \dfrac{1}{|v|} & \text{if } v \neq 0 \\ 0 & \text{if } v = 0. \end{cases}$$

Recall that we wanted $f_U(u)$, whence

$$f_U(u) = \int_{-\infty}^{\infty} f_{X,Y}(v, u/v) \frac{1}{|v|} dv.$$

3. Sometimes we can calculate a density directly, without resorting to Theorem 12.6 and (for example) the trick of Example 2 above.
   Let X, Y be independent and both be normal with parameters $\mu = 0$ and $\sigma^2 = 1$. Let $Z = X^2 + Y^2$. What is $f_Z(z)$?
   By Theorem 12.2(b) the density of the pair (X, Y) is

$$f(x, y) = \frac{1}{2\pi} \exp\left(-\frac{x^2 + y^2}{2}\right),$$

and therefore

$$P(Z \in A) = E\{1_A(Z)\} = E\{1_A(X^2 + Y^2)\}$$

$$= \int\!\!\int 1_A(x^2 + y^2) f(x, y) dx \, dy.$$

$$= \int\!\!\int 1_A(x^2 + y^2) \frac{1}{2\pi} e^{-(\frac{x^2+y^2}{2})} dx \, dy$$

By changing to polar coordinates we have

$$= \frac{1}{2\pi} \int_0^{\infty} \int_0^{2\pi} 1_A(r^2) e^{-r^2/2} \, r \, dr \, d\theta$$

$$= \frac{1}{2\pi} \int_0^{2\pi} d\theta \int_0^{\infty} 1_A(r^2) e^{-r^2/2} \, r \, dr$$

$$= \int_0^{\infty} 1_A(r^2) e^{-r^2/2} \, r \, dr.$$

Now let $z = r^2$; then $dz = 2r \, dr$:

$$= \int_0^{\infty} 1_A(z) \frac{1}{2} e^{-(z/2)} dz,$$

and we see that Z has the density $\frac{1}{2} e^{-(z/2)}$ of an Exponential r.v. with parameter $\frac{1}{2}$. Note that the polar coordinates transformation is not bijective from $\mathbf{R}^2$ into its range, so to straighten out the above argument we have to resort again to Corollary 12.1: this transformation is bijective from $S_1 = \mathbf{R}^2\backslash\{0\}$ into $(0, \infty) \times (0, 2\pi]$, while the set $S_0 = \{0\}$ is of Lebesgue measure zero. This argument will be used without further notice in the sequel.

4. When a function g transforms n random variables to one, we can sometimes avoid combining Theorem 12.6 with the trick of Example 2, known as auxiliary random variables, by using instead the distribution function. More specifically, if $Y = g(X_1, \ldots, X_n)$ and if f is the joint density of $(X_1, \ldots, X_n)$, then

$$F_Y(y) = P(Y \le y) = \int_{g(x_1,\ldots,x_n) \le y} f(x)dx.$$

Suppose there exists a function $h(y; x_2, \ldots, x_n)$ such that

$$g(x_1, x_2, \ldots, x_n) \le y \text{ if and only if } x_1 \le h(y; x_2, \ldots, x_n).$$

Then

$$F_Y(y) = \int_{\{x:g(x) \le y\}} f(x)dx$$

$$= \int_{-\infty}^{\infty} \cdots \int_{-\infty}^{\infty} \left\{ \int_{-\infty}^{h(y;x_2,\ldots,x_n)} f(x_1,\ldots,x_n)dx_1 \right\} dx_2, \ldots, dx_n.$$

If we differentiate both sides with respect to y and assume that h is continuously differentiable in y and f is continuous, then we get

$$f_Y(y) = \frac{d}{dy} F_Y(y) \qquad (12.3)$$

$$= \int_{-\infty}^{\infty} \cdots \int_{-\infty}^{\infty} \frac{\partial h(y; x_2, \ldots, x_n)}{\partial y}$$

$$\times f\left(h(y, x_2, \ldots, x_n), x_2, \ldots, x_n\right) dx_2 \ldots dx_n.$$

An example of this technique which is useful in deriving distributions important in statistical inference is as follows: Let X, Y be independent r.v.'s, and let X be normal ($\mu = 0, \sigma^2 = 1$) and let Y be Gamma ($\alpha = n/2; \beta = \frac{1}{2}$; assume $n \in \mathbf{N}$). [Note: Y is also the *"chi-square with n degrees of freedom"*.] What is the distribution of

$$Z = \frac{X}{\sqrt{Y/n}} \; ?$$

Here $g(x, y) = \frac{x}{\sqrt{y/n}}$, and since $\frac{x}{\sqrt{y/n}} \le z$ if and only if $x \le z\sqrt{y/n}$, we have $h(z; y) = z\sqrt{y/n}$. By independence we have that the joint density f of (X, Y) is

$$f(x, y) = f_X(x) f_Y(y) = \left(\frac{1}{\sqrt{2\pi}} e^{-\frac{x^2}{2}} \right) \left(\frac{y^{\frac{n}{2}-1} e^{-\frac{1}{2}y}}{2^{\frac{n}{2}} \Gamma(\frac{n}{2})} \right)$$

for $-\infty < x < \infty$ and $0 < y < \infty$, and $f(x, y) = 0$ otherwise. We can now apply (12.3) to obtain

$$f_Z(z) = \frac{1}{\sqrt{2\pi n} \Gamma(\frac{n}{2}) 2^{\frac{n}{2}}} \int_0^{\infty} y^{\frac{n-1}{2}} e^{-y \frac{(1+\frac{z^2}{n})}{2}} dy \qquad (12.4)$$

$$= \frac{\Gamma(\frac{n+1}{2})}{\Gamma(\frac{n}{2})\sqrt{\pi n}(1 + \frac{z^2}{n})^{\frac{1}{2}(n+1)}}, \qquad -\infty < z < \infty.$$

We note that the density of Z in (12.4) above is called the density for the *Student's t-distribution with n degrees of freedom*.

The Student's t-distribution was originally derived as the distribution of a random variable arising from statistical inference for the mean of a normal distribution. It was originally studied by W. Gosset (1876–1937)writing under the pseudonym "Student".

5. Let X, Y be independent normal r.v.'s with mean $\mu = 0$ and variance $\sigma^2 < \infty$. Let $Z = \sqrt{X^2 + Y^2}$ and $W = \frac{X}{Y}$ if $Y \neq 0$ and $W = 0$ if $Y = 0$. We wish to find $f_{(Z,W)}(z, w)$, the joint density of (Z, W). Here

$$g(x, y) = \left(\sqrt{x^2 + y^2}, \frac{x}{y} \right) = (z, w)$$

and g is not injective. We have

$$g^{-1}(z, w) = \left(\frac{zw}{\sqrt{1 + w^2}}, \frac{z}{\sqrt{1 + w^2}} \right) = (x, y).$$

A second inverse would be

$$h^{-1}(z, w) = \left(\frac{-zw}{\sqrt{1 + w^2}}, \frac{-z}{\sqrt{1 + w^2}} \right).$$

The Jacobian $J_{g^{-1}}$ is given by

$$\begin{pmatrix} \dfrac{w}{\sqrt{1 + w^2}} & \dfrac{1}{\sqrt{1 + w^2}} \\ \dfrac{z}{(1 + w^2)^{\frac{3}{2}}} & \dfrac{-zw}{(1 + w^2)^{\frac{3}{2}}} \end{pmatrix}$$

and its determinant is $\frac{-z}{1+w^2}$. Therefore by Corollary 12.1 we have

$$f_{(Z,W)}(z, w) = \frac{z}{1 + w^2} \left\{ f_{(X,Y)} \left(\frac{zw}{\sqrt{1 + w^2}}, \frac{z}{\sqrt{1 + w^2}} \right) \right.$$
$$\left. + f_{(X,Y)} \left(\frac{-zw}{\sqrt{1 + w^2}}, \frac{-z}{\sqrt{1 + w^2}} \right) \right\}.$$

In this case the normal density is symmetric:

$$f_{(X,Y)}(x, y) = f_{(X,Y)}(-x, -y),$$

hence we get

$$f_{(Z,W)}(z, w) = \frac{2z}{1 + w^2} \frac{1}{2\pi\sigma^2} e^{-\frac{z^2}{2\sigma^2}} 1_{(z>0)}.$$

Note that the density factors:

$$f_{(Z,W)}(z, w) = \frac{1}{\pi \sigma^2} \left(\frac{1}{1 + w^2} \right) \left(z e^{-\frac{z^2}{2\sigma^2}} 1_{(z>0)} \right).$$

Therefore we deduce that Z and W are independent (which is not *a priori* obvious), and we can even read off the densities of Z and W if we can infer the normalizing constants from each. Indeed, since $\frac{1}{\pi(1+w^2)}$ is the density of a Cauchy random variable (with $\alpha = 0$ and $\beta = 1$), we conclude:

$$f_Z(z) = \frac{z}{\sigma^2} e^{-\frac{z^2}{2\sigma^2}} 1_{(z>0)}$$

and

$$f_W(w) = \frac{1}{\pi(1 + w^2)}, \qquad -\infty < w < \infty,$$

and Z and W are independent. The distribution density f_Z above is known as the *Rayleigh density with parameter* $\sigma^2 > 0$. This example shows also that the *ratio of two independent normals with mean 0 is a Cauchy r.v.* ($\alpha = 0$ and $\beta = 1$).

Exercises for Chapter 12

12.1 Show that

$$\int_{-\infty}^{\infty} \int_{-\infty}^{\infty} e^{-\frac{(x^2+y^2)}{2\sigma^2}} \, dx \, dy = 2\pi\sigma^2,$$

and therefore that $\frac{1}{2\pi\sigma^2} e^{-(x^2+y^2)/2\sigma^2}$ is a true density. (*Hint:* Use polar coordinates.)

12.2 Suppose a joint density $f_{(X,Y)}(x,y)$ factors: $f_{(X,Y)}(x,y) = g(x)h(y)$. Find $f_X(x)$ and $f_Y(y)$.

12.3 Let (X,Y) have joint density

$$f(x,y) = \frac{1}{2\pi\sigma_1\sigma_2\sqrt{1-r^2}}$$
$$\times \exp\left(-\frac{1}{2(1-r^2)}\left\{\frac{(x-\mu_1)^2}{\sigma_1^2} - \frac{2r(x-\mu_1)(y-\mu_2)}{\sigma_1\sigma_2} + \frac{(y-\mu_2)^2}{\sigma_2^2}\right\}\right).$$

Find $f_{X=x}(y)$. [Ans: $\frac{1}{\sigma_2\sqrt{2\pi(1-r^2)}} \exp(-\frac{1}{2\sigma_2^2(1-r^2)}\{y-\mu_2 - \frac{r\sigma_2}{\sigma_1}(x-\mu_1)\}^2).$]

12.4 Let $\rho_{X,Y}$ denote the correlation coefficient for (X,Y). Let $a > 0$, $c > 0$ and $b \in \mathbf{R}$. Show that

$$\rho_{aX+b,cY+b} = \rho_{X,Y}.$$

(This is useful since it shows that ρ is independent of the scale of measurement for X and Y.)

12.5 If $a \neq 0$, show that

$$\rho_{X,aX+b} = \frac{a}{|a|},$$

so that if $Y = aX+b$ is an affine non-constant function of X, then $\rho_{X,Y} = \pm 1$.

12.6 Let X, Y have finite variances and let

$$Z = \left(\frac{1}{\sigma_Y}\right)Y - \left(\frac{\rho_{X,Y}}{\sigma_X}\right)X.$$

Show that $\sigma_Z^2 = 1 - \rho_{X,Y}^2$, and deduce that if $\rho_{X,Y} = \pm 1$, then Y is a non-constant affine function of X.

12.7 * (Gut (1995), p. 27.) Let (X,Y) be *uniform on the unit ball:* that is,

$$f_{(X,Y)}(x,y) = \begin{cases} \dfrac{1}{\pi} & \text{if } x^2 + y^2 \leq 1 \\ 0 & \text{if } x^2 + y^2 > 1. \end{cases}$$

Find the distribution of $R = \sqrt{X^2 + Y^2}$. (*Hint:* Introduce an auxiliary r.v. $S = \text{Arctan}\left(\frac{Y}{X}\right)$.) [Ans: $f_R(r) = 2r1_{(0,1)}(r).$]

12.8 Let (X, Y) have density $f(x, y)$. Find the density of $Z = X + Y$. (*Hint:* Find the joint density of (Z, W) first where $W = Y$.) [Ans: $f_Z(z) = \int_{-\infty}^{\infty} f_{(X,Y)}(z - w, w) dw$.]

12.9 Let X be normal with $\mu = 0$ and $\sigma^2 < \infty$, and let Θ be uniform on $(0, \pi)$: that is $f(\theta) = \frac{1}{\pi} 1_{(0,\pi)}(\theta)$. Assume X and Θ are independent. Find the distribution of $Z = X + a \cos(\Theta)$. (This is useful in electrical engineering.) [Ans: $f_Z(z) = \frac{1}{\pi \sigma \sqrt{2\pi}} \int_0^\pi e^{-(z - a \cos w)^2 / 2\sigma^2} dw$.]

12.10 Let X and Y be independent and suppose $Z = g(X)$ and $W = h(Y)$, with g and h both injective and differentiable. Find a formula for $f_{Z,W}(z, w)$, the joint density of (Z, W).

12.11 Let (X, Y) be independent normals, both with means $\mu = 0$ and variances σ^2. Let

$$ Z = \sqrt{X^2 + Y^2} \text{ and } W = \text{Arctan}\left(\frac{X}{Y}\right), \qquad -\frac{\pi}{2} < W \le \frac{\pi}{2}. $$

Show that Z has a Rayleigh distribution, that W is uniform on $(-\frac{\pi}{2}, \frac{\pi}{2})$, and that Z and W are independent.

12.12 Let $(X_1, \ldots, X_n)$ be random variables. Define

$$ Y_1 = \min(X_i; 1 \le i \le n) $$
$$ Y_2 = \text{ second smallest of } X_1, \ldots, X_n $$
$$ \vdots $$
$$ Y_n = \text{ largest of } X_1, \ldots, X_n. $$

Then $Y_1, \ldots, Y_n$ are also random variables, and $Y_1 \le Y_2 \le \ldots \le Y_n$. Thus the Y random variables are the same as the X ones, but they are arranged *in order*. They are called the *order statistics* of $(X_1, \ldots, X_n)$ and are usually denoted

$$ Y_k = X_{(k)}. $$

Assume the X_i are i.i.d. with common density f. Show that the joint density of the order statistics is given by

$$ f_{(X_{(1)}, \ldots, X_{(n)})}(y_1, \ldots, y_n) = \begin{cases} n! \prod_{i=1}^n f(y_i) & \text{for } y_1 < y_2 < \ldots < y_n \\ 0 & \text{otherwise .} \end{cases} $$

12.13 Let $(X_1, \ldots, X_n)$ be i.i.d. uniform on $(0, a)$. Show that the order statistics $(X_{(1)}, \ldots, X_{(n)})$ have density

$$ f(y_1, \ldots, y_n) = \begin{cases} \frac{n!}{a^n} & \text{for } y_1 < y_2 < \ldots < y_n \\ 0 & \text{otherwise .} \end{cases} $$

(*Hint:* Use Exercise 12.12.)

12.14 Show that if $(X_1, \ldots, X_n)$ are i.i.d. with common density f and distribution function F, then $X_{(k)}$ (see Exercise 12.12) has density

$$f_{(k)}(y) = k \binom{n}{k} f(y)(1 - F(y))^{n-k} F(y)^{k-1}.$$

12.15 (Simulation of Normal Random Variables.) Let U_1, U_2 be two independent uniform random variables on $(0, 1)$. Let $\theta = 2\pi U_1$ and let $S = -\ln(U_2)$.

a) Show that S has an exponential distribution, and that $R = \sqrt{2S}$ has a Rayleigh distribution.
b) Let $X = R \cos \theta$, $Y = R \sin \theta$. Show that X and Y are independent normals.

(*Hint:* For part (a), recall that an exponential is a special case of a Gamma distribution: indeed, it is χ_2^2. Then for part (b) reverse the procedure of Exercise 12.11.)

Remark: Exercise 12.15 is known as the Box–Muller method for simulating normal random variables.

13 Characteristic Functions

It often arises in mathematics that one can solve problems and/or obtain properties of mathematical objects by "transforming" them into another space, solving the problem there, and then transforming the solution back. Two of the most important transforms are the Laplace transform and the Fourier transform. While these transforms are widely used in the study of differential equations, they are also extraordinarily useful for the study of Probability. They can be used to analyze random variables (e.g., to compute their moments), and they can be used to give short and elegant proofs of the Central Limit Theorem (see Chapter 21). The Fourier transform is the more sophisticated of the two, and it is also the most useful.

Let us write $\langle x, y \rangle$ for the scalar product of $x, y \in \mathbf{R}^n$. That is, if $x = (x_1, \ldots, x_n)$ and $y = (y_1, \ldots, y_n)$, then

$$\langle x, y \rangle = \sum_{j=1}^{n} x_j y_j.$$

(This is often written $x \cdot y$ and called the "dot product" in Calculus courses.)

Definition 13.1. *Let μ be a probability measure on $\mathbf{R}^n$. Its Fourier transform is denoted $\hat{\mu}$ and is a function on $\mathbf{R}^n$ given by*

$$\hat{\mu}(u) = \int e^{i\langle u, x \rangle} \mu(dx).$$

In the above definition i denotes the square root of negative one ($i = \sqrt{-1}$). We integrate the complex-valued function $x \to e^{i\langle u, x \rangle}$; however, no difficulty is involved here, since we have

$$e^{i\langle u, x \rangle} = \cos(\langle u, x \rangle) + i \sin(\langle u, x \rangle), \tag{13.1}$$

and in particular $|e^{iux}| = 1$. (The equation (13.1) can be verified in an elementary way by using the power series expansions of e^z, $\cos z$, and $\sin z$.) Now, both functions $x \to \cos(\langle u, x \rangle)$ and $x \to \sin(\langle u, x \rangle)$ are bounded and Borel, hence integrable, and the formula of Definition 13.1 becomes

$$\hat{\mu}(u) = \int \cos(\langle u, x \rangle) \mu(dx) + i \int \sin(\langle u, x \rangle) \mu(dx).$$

As a matter of fact, all results of Chapter 9 hold for complex-valued functions just by taking separately the real and imaginary parts. These results will be used without further mention in the sequel (the only result which is more subtle to get in the complex case is that $|\int f d\mu| \leq \int |f| d\mu$ for the modulus).

Definition 13.2. *Let X be an $\mathbf{R}^n$-valued random variable. Its characteristic function φ_X defined on $\mathbf{R}^n$ is*

$$\varphi_X(u) = E\{e^{i\langle u, X\rangle}\}.$$

We note that

$$\varphi_X(u) = \int e^{i\langle u, x\rangle} P^X(dx) = \widehat{P^X}(u) \tag{13.2}$$

where P^X is the distribution measure of X. Therefore characteristic functions always exist because they are equal to Fourier transforms of probability measures, which we have just seen always exist.

Theorem 13.1. *Let μ be a probability measure on $\mathbf{R}^n$. Then $\hat{\mu}$ is a bounded, continuous function with $\hat{\mu}(0) = 1$.*

Proof. We have already seen that $\hat{\mu}$ always exists; that is, its definition always makes sense. Since $|e^{i\langle u, x\rangle}| = 1$ for all real u, x, we have

$$|\hat{\mu}(u)| \leq \int |e^{i\langle u, x\rangle}| \mu(dx) = \int 1 \mu(dx) = 1.$$

Moreover

$$\hat{\mu}(0) = \int e^{i\langle 0, x\rangle} \mu(dx) = \int 1 \mu(dx) = 1.$$

Finally suppose u_p tends to u; we wish to show $\hat{\mu}(u_p)$ tends to $\hat{\mu}(u)$. The functions $e^{i\langle u_p, x\rangle}$ converge pointwise to $e^{i\langle u, x\rangle}$, and since $x \to e^{i\langle u, x\rangle}$ is bounded in modulus by 1, the continuity follows from Lebesgue's dominated convergence theorem (Theorem 9.1(f)). □

Actually one can show that $\hat{\mu}$ is *uniformly* continuous, but we do not need such a result here.

Theorem 13.2. *Let X be an $\mathbf{R}^n$ valued random variable and suppose $E\{|X|^m\} < \infty$ for some integer m. Then the characteristic function φ_X of X has continuous partial derivatives up to order m, and*

$$\frac{\partial^m}{\partial x_{j_1} \dots \partial x_{j_m}} \varphi_X(u) = i^m E\{X_{j_1} \dots X_{j_m} e^{i\langle u, X\rangle}\}.$$

Proof. We prove an equivalent formulation stated in terms of Fourier transforms of probability measures. (To see the equivalence, simply take μ to be P^X, the distribution measure on $\mathbf{R}^n$ of X as in (13.2).) Let μ be a probability measure on $\mathbf{R}^n$ and assume $f(x) = |x|^m$ is integrable:

$$\int_{\mathbf{R}^n} |x|^m \mu(dx) < \infty.$$

Then we wish to show that $\hat{\mu}(u)$ is m-times continuously differentiable and

$$\frac{\partial^m \hat{\mu}}{\partial x_{j_1} \dots \partial x_{j_m}}(u) = i^m \int x_{j_1} \dots x_{j_m} e^{i\langle u, x\rangle} \mu(dx).$$

We give the proof only for the case $m = 1$. The general case can be established analogously by recurrence.

In order to prove that $\frac{\partial \hat{\mu}}{\partial x_j}$ exists at point u, it is enough to prove that for every sequence of reals t_p tending to 0, and with $v = (v_1, \dots, v_n)$ being the unit vector in $\mathbf{R}^n$ in the direction j (i.e. with coordinates $v_k = 0$ for $k \neq j$ and $v_j = 1$), then the sequence

$$\frac{1}{t_p}\{\hat{\mu}(u + t_p v) - \hat{\mu}(u)\} = \int e^{i\langle u, x\rangle} \frac{e^{it_p x_j} - 1}{t_p} \mu(dx). \qquad (13.3)$$

converges to a limit independent of the sequence t_p, and in this case this limit equals $\frac{\partial \hat{\mu}}{\partial x_j}(u)$. The sequence of functions $x \to \frac{e^{it_p x_j} - 1}{t}$ (where x_j is the jth coordinate of $x \in \mathbf{R}^n$) converges pointwise to ix_j by differentiation; moreover

$$\left| \frac{e^{it_p x_j} - 1}{t_p} \right| \leq 2|x|,$$

and

$$\int 2|x| \mu(dx) < \infty$$

by hypothesis. Therefore by Lebesgue's dominated convergence theorem (Theorem 9.1(f)) we have that (13.3) converges to

$$i \int x_j e^{i\langle u, x\rangle} \mu(dx).$$

Therefore

$$\frac{\partial \hat{\mu}(u)}{\partial x_j} = i \int e^{i\langle u, x\rangle} x_j \mu(dx). \qquad (13.4)$$

The proof that the partial derivative in (13.4) above is continuous is exactly the same as that of Theorem 13.1.   □

An immediate application of the above is to use characteristic functions to calculate the moments of random variables. (The kth *moment* of a r.v. X is $E\{X^k\}$.) For the first two moments (by far the most important) we note that for X real valued (by Theorem 13.2):

$$E\{X\} = -i\varphi_X'(0) \text{ if } E\{|X|\} < \infty \qquad (13.5)$$
$$E\{X^2\} = -\varphi_X''(0) \text{ if } E\{X^2\} < \infty. \qquad (13.6)$$

Examples:

1. *Bernoulli (p):* If X is Bernoulli with parameter p, then

$$\varphi_X(u) = E\{e^{iuX}\} = e^{iu0}(1-p) + e^{iu}p = \boxed{pe^{iu} + 1 - p}$$

2. *Binomial $B(p,n)$:* If X is Binomial with parameters n, p, then

$$\varphi_X(u) = E\{e^{iuX}\} = \sum_{j=0}^{n}\binom{n}{j}e^{iuj}p^j(1-p)^{n-j} = \boxed{(pe^{iu}+1-p)^n}.$$

We could also have noted that

$$X = \sum_{j=1}^{n}Y_j,$$

where $Y_1, \ldots, Y_n$ are independent and Bernoulli (p). Then

$$\varphi_X(u) = E\{e^{iuX}\} = E\{e^{iu\sum_{j=1}^{n}Y_j}\} = E\left\{\prod_{j=1}^{n}e^{iuY_j}\right\} = \prod_{j=1}^{n}E\{e^{iuY_j}\}$$

by the independence of the Y_j's;

$$= \prod_{j=1}^{n}\varphi_{Y_j}(u) = (pe^{iu}+1-p)^n.$$

3. *Poisson (λ):*

$$\varphi_X(u) = E\{e^{iuX}\} = \sum_{k=0}^{\infty}e^{iuk}P(X=k)$$

$$= \sum_{k=0}^{\infty}e^{iuk}\frac{\lambda^k}{k!}e^{-\lambda} = \sum_{k=0}^{\infty}\frac{(\lambda e^{iu})^k}{k!}e^{-\lambda}$$

$$= e^{-\lambda}e^{\lambda e^{iu}} = \boxed{e^{\lambda(e^{iu}-1)}}.$$

4. *Uniform on $(-a, a)$:*

$$\varphi_X(u) = E\{e^{iuX}\} = \frac{1}{2a}\int_{-a}^{a}e^{iux}dx = \frac{e^{iua} - e^{-iua}}{2aiu};$$

using that $e^z = \cos z + i\sin z$, and that $\cos(a) = \cos(-a)$, this equals

$$= \frac{2i\sin au}{2aiu} = \boxed{\frac{\sin au}{au}}.$$

5. *The Normal* ($\mu = 0; \sigma^2 = 1$): Calculating the characteristic function of the normal is a bit hard. It can be done via contour integrals and the residue theorem (using the theory of complex variables), or by analytic continuation (see Exercise 17 of Chapter 14); we present here a perhaps non-intuitive method that has the virtue of being elementary:

$$\varphi_X(u) = \int e^{iux} \frac{1}{\sqrt{2\pi}} e^{-x^2/2} dx$$

$$= \int \frac{\cos ux}{\sqrt{2\pi}} e^{-x^2/2} dx + i \int \frac{\sin ux}{\sqrt{2\pi}} e^{-x^2/2} dx.$$

Since $\sin ux\, e^{-x^2/2}$ is an odd and integrable function, we have that

$$\int_{-\infty}^{\infty} \sin ux\, e^{-x^2/2} dx = 0,$$

and thus

$$\varphi_X(u) = \frac{1}{\sqrt{2\pi}} \int_{-\infty}^{\infty} \cos ux\, e^{-x^2/2} dx.$$

By Theorem 13.2 we can differentiate both sides with respect to u to obtain:

$$\varphi_X'(u) = \frac{1}{\sqrt{2\pi}} \int_{-\infty}^{\infty} -x \sin ux\, e^{-x^2/2} dx.$$

Next integrate by parts to get:

$$= -\frac{1}{\sqrt{2\pi}} \int_{-\infty}^{\infty} u \cos ux\, e^{-x^2/2} dx = -u\varphi_X(u).$$

This gives us the ordinary differential equation

$$\frac{\varphi_X'}{\varphi_X} = -u,$$

and anti-differentiating both sides yields

$$\ln|\varphi_X(u)| = -\frac{u^2}{2} + C,$$

and exponentiating gives

$$\varphi_X(u) = e^C e^{-u^2/2}.$$

Since $\varphi_X(0) = 1$, we deduce $e^C = 1$, whence

$$\boxed{\varphi_X(u) = e^{-u^2/2}}.$$

Theorem 13.3. *Let X be an $\mathbf{R}^n$-valued random variable and $a \in \mathbf{R}^m$. Let A be an $m \times n$ matrix. Then*

$$\varphi_{a+AX}(u) = e^{i\langle u,a \rangle} \varphi_X(A^* u),$$

for all $u \in \mathbf{R}^m$, where A^* denotes A transpose.

Proof. One easily verifies that

$$e^{i\langle u, a+AX \rangle} = e^{i\langle u,a \rangle} e^{i\langle A^* u, X \rangle},$$

and then taking expectations of both sides gives the result.   □

Examples (continued)

6. *The Normal (μ, σ^2):* Let X be $N(\mu, \sigma^2)$. Then one easily checks (see Exercise 14.18) that $Y = \frac{X-\mu}{\sigma}$ is Normal $(0, 1)$. Alternatively, X can be written $X = \mu + \sigma Y$, where Y is $N(0, 1)$. Then using Theorem 13.3 and example 5 we have

$$\boxed{\varphi_X = e^{iu\mu - u^2 \sigma^2 / 2}}$$

7. *The Exponential (λ):* Let X be Exponential with parameter λ. Then

$$\varphi_X(u) = \int_0^\infty e^{iux} \lambda e^{-\lambda x} dx.$$

A formal calculation gives

$$= \int_0^\infty \lambda e^{(iu-\lambda)x} dx = \boxed{\frac{\lambda}{\lambda - iu}}$$

but this is not mathematically rigorous. It can be justified by, for example, a contour integral using complex analysis. Another method is as follows: It is easy to check that the functions

$$\frac{\lambda}{\lambda^2 + u^2} e^{-\lambda x} (-\lambda \cos(ux) + u \sin(ux)),$$

$$\frac{\lambda}{\lambda^2 + u^2} e^{-\lambda x} (-u \cos(ux) - \lambda \sin(ux)),$$

have derivatives $\lambda e^{-\lambda x} \cos(ux)$ and $\lambda e^{-\lambda x} \sin(ux)$ respectively. Thus

$$\int_0^\infty \lambda e^{-\lambda x} \cos(ux) dx = \frac{\lambda}{\lambda^2 + u^2} e^{-\lambda x} (-\lambda \cos(ux) + u \sin(ux)) \Big|_0^\infty,$$

$$\int_0^\infty \lambda e^{-\lambda x} \sin(ux) dx = \frac{\lambda}{\lambda^2 + u^2} e^{-\lambda x} (-u \cos(ux) - \lambda \sin(ux)) \Big|_0^\infty.$$

Hence we get

$$\varphi_X(u) = \frac{\lambda^2}{\lambda^2 + u^2} - i \frac{\lambda u}{\lambda^2 + u^2} = \frac{\lambda}{\lambda - iu}.$$

8. *The Gamma* (α, β): One can show using contour integration and the residue theorem in the theory of complex variables that if X is Gamma (α, β) then

$$\varphi_X(u) = \frac{\beta^\alpha}{(\beta - iu)^\alpha}.$$

One can also calculate the characteristic function of a Gamma random variable without resorting to contour integration: see Exercise 14.19.

14 Properties of Characteristic Functions

We have seen several examples on how to calculate a characteristic function when given a random variable. Equivalently we have seen examples of how to calculate the Fourier transforms of probability measures. For such transforms to be useful, we need to know that knowledge of the transform characterizes the distribution that gives rise to it. The proof of the next theorem uses the Stone-Weierstrass theorem and thus is a bit advanced for this book. Nevertheless we include the proof for the sake of completeness.

Theorem 14.1 (Uniqueness Theorem). *The Fourier transform $\hat{\mu}$ of a probability measure μ on $\mathbf{R}^n$ characterizes μ: that is, if two probabilities on $\mathbf{R}^n$ admit the same Fourier transform, they are equal.*

Proof. Let

$$f(\sigma, x) = \frac{1}{(2\pi\sigma^2)^{n/2}} e^{-|x|^2/2\sigma^2},$$

and

$$\hat{f}(\sigma, u) = e^{-|u|^2\sigma^2/2}.$$

Then $f(\sigma, x)$ is the density of $X = (X_1, \ldots, X_n)$, where the X_j's are independent and $N(0, \sigma^2)$ for each j $(1 \leq j \leq n)$. By Example 6 of Chapter 13 and the Tonelli-Fubini Theorem, we have

$$\int_{\mathbf{R}^n} f(\sigma, x) e^{i\langle u, x\rangle} \, dx = \int_{\mathbf{R}^n} \prod_{j=1}^{n} \frac{1}{\sigma\sqrt{2\pi}} \, e^{\left(\frac{-x_j^2}{2\sigma^2} + iu_j x_j\right)} dx_1 \ldots dx_n$$

$$= \prod_{j=1}^{n} \int_{\mathbf{R}} \frac{1}{\sigma\sqrt{2\pi}} \, e^{\left(\frac{-x_j^2}{2\sigma^2} + iu_j x_j\right)} dx_j$$

$$= \prod_{j=1}^{n} e^{-\frac{u_j^2\sigma^2}{2}} = \hat{f}(\sigma, u).$$

Therefore

$$f(\sigma, u - v) = \frac{1}{(2\pi\sigma^2)^{n/2}} \hat{f}\left(\sigma, \frac{u-v}{\sigma^2}\right)$$

$$= \frac{1}{(2\pi\sigma^2)^{n/2}} \int_{\mathbf{R}^n} f(\sigma, x) e^{i\langle \frac{u-v}{\sigma^2}, x\rangle} dx.$$

Next suppose that μ_1 and μ_2 are two probability measures on $\mathbf{R}^n$ with the same Fourier transforms $\hat{\mu}_1 = \hat{\mu}_2 = \hat{\mu}$. Then

$$\int f(\sigma, u - v)\mu_1(du)$$

$$= \int \frac{1}{(2\pi\sigma^2)^{n/2}} \left\{ \int f(\sigma, x) e^{i\langle \frac{u-v}{\sigma^2}, x \rangle} dx \right\} \mu_1(du)$$

$$= \int f(\sigma, x) \frac{1}{(2\pi\sigma^2)^{n/2}} \hat{\mu}\left(\frac{x}{\sigma^2}\right) e^{-i\frac{\langle v, x \rangle}{\sigma^2}} dx,$$

(the reader will check that one can apply Fubini's theorem here), and the exact same equalities hold for μ_2. We conclude that

$$\int g(x)\mu_1(dx) = \int g(x)\mu_2(dx)$$

for all $g \in \mathcal{H}$, where $\mathcal{H}$ is the vector space generated by all functions of the form $u \to f(\sigma, u - v)$. We then can apply the Stone-Weierstrass theorem[1] to conclude that $\mathcal{H}$ is dense in $\mathcal{C}_0$ under uniform convergence, where $\mathcal{C}_0$ is the set of functions "vanishing at ∞": that is, $\mathcal{C}_0$ consists of all continuous functions on $\mathbf{R}^n$ such that $\lim_{\|x\| \to \infty} |f(x)| = 0$. We then obtain that

$$\int_{\mathbf{R}^n} g(x)\mu_1(dx) = \int_{\mathbf{R}^n} g(x)\mu_2(dx)$$

for all $g \in \mathcal{C}_0$. Since the indicator function of an open set can be written as the increasing limit of functions in $\mathcal{C}_0$, the Monotone Convergence Theorem (Theorem 9.1(d)) then gives

$$\mu_1(A) = \mu_2(A), \quad \text{all open sets } A \subset \mathbf{R}^n.$$

Finally the Monotone Class Theorem (Theorem 6.2) gives

$$\mu_1(A) = \mu_2(A) \quad \text{for all Borel sets } A \subset \mathbf{R}^n,$$

which means $\mu_1 = \mu_2$. □

Corollary 14.1. *Let $X = (X_1, \ldots, X_n)$ be an $\mathbf{R}^n$-valued random variable. Then the real-valued r.v.'s $(X_j)_{1 \leq j \leq n}$ are independent if and only if*

$$\varphi_X(u_1, \ldots, u_n) = \prod_{j=1}^{n} \varphi_{X_j}(u_j) \tag{14.1}$$

---

[1] One can find a nice treatment of the Stone–Weierstrass theorem in [20, p. 160].

Proof. If the $(X_j)_{1 \le j \le n}$ are independent, then

$$\varphi_X(u) = E\{e^{i\langle u, X \rangle}\} = E\left\{e^{i \sum_{j=1}^n u_j X_j}\right\}$$

$$= E\left\{\prod_{j=1}^n e^{iu_j X_j}\right\} = \prod_{j=1}^n E\{e^{iu_j X_j}\}$$

by the independence;

$$= \prod_{j=1}^n \varphi_{X_j}(u_j).$$

Next suppose we have (14.1). Let μ_X denote the law of X on $\mathbf{R}^n$ and let μ_{X_j} denote the law of X_j on $\mathbf{R}$. Then

$$\hat{\mu}_X = (\mu_{X_1} \otimes \mu_{X_2} \otimes \ldots \otimes \mu_{X_n})^{\hat{}},$$

and therefore by Theorem 14.1 we have

$$\mu_X = \mu_{X_1} \otimes \mu_{X_2} \otimes \ldots \otimes \mu_{X_n},$$

which is equivalent to independence. □

Caution: In the above, having $\varphi_X(u, u, \ldots, u) = \prod_{i=1}^n \varphi_{X_j}(u)$ for all $u \in \mathbf{R}$ is *not* enough for the r.v.'s X_j to be independent.

Exercises for Chapters 13 and 14

Note: The first three exercises require the use of contour integration and the residue theorem from complex analysis. These problems are given the symbol "♯".

14.1 ♯ Let $f(x) = \frac{1}{\pi(1+x^2)}$, a Cauchy density, for a r.v. X. Show that

$$\varphi_X(u) = e^{-|u|},$$

by integrating around a semicircle with diameter $[-R, R]$ on the real axis to the left, from the point $(-R, 0)$ to the point $(R, 0)$ over the real axis.

14.2 ♯ Let X be a gamma r.v. with parameters (α, β). Show using contour integration that $\varphi_X(u) = \frac{\beta^\alpha}{(\beta-iu)^\alpha}$. [*Hint:* Use the contour for $0 < c < d$ on the real axis, go from $(d, 0)$ back to $(c, 0)$, then descend vertically to the line $y = -\frac{ux}{\alpha}$ and descend southeast along the line, and then ascend vertically to $(d, 0)$.]

14.3 ♯ Let X be $N(0, 1)$ (i.e., normal with $\mu = 0$ and $\sigma^2 = 1$), and show that $\varphi_X(u) = e^{-u^2/2}$ by contour integration. [*Hint:* use the contour from $(R, 0)$ to $(-R, 0)$ on the real axis; then descend vertically to $(-R, -iu)$; then proceed horizontally to $(R, -iu)$, and then ascend vertically back to the real axis.]

14.4 * Suppose $E\{|X|^2\} < \infty$ and $E\{X\} = 0$. Show that $\text{Var}(X) = \sigma^2 < \infty$, and that

$$\varphi_X(u) = 1 - \frac{1}{2}u^2\sigma^2 + o(u^2)$$

as $u \to 0$. [Recall that a function g is $o(t)$ if $\lim_{t \to 0} \frac{|g(t)|}{t} = 0$.]

14.5 Let $X = (X_1, \ldots, X_n)$ be an $\mathbf{R}^n$ valued r.v. Show that

a) $\varphi_X(u, 0, 0, \ldots, 0) = \varphi_{X_1}(u)$ $(u \in \mathbf{R})$
b) $\varphi_X(u, u, u, \ldots, u) = \varphi_{X_1 + \ldots + X_n}(u)$ $(u \in \mathbf{R})$

14.6 Let $\bar{z}$ denote the complex conjugate of z. That is, if $z = a + ib$ then $\bar{z} = a - ib$ $(a, b \in \mathbf{R})$. Show that for X a r.v.,

$$\overline{\varphi_X(u)} = \varphi_X(-u).$$

14.7 Let X be a r.v. Show that $\varphi_X(u)$ is a real-valued function (as opposed to a complex-valued function) if and only if X has a symmetric distribution. (That is, $P^X = P^{-X}$, where P^X is the distribution measure of X.) [*Hint:* Use Exercise 14.6, Theorem 13.3, and Theorem 14.1.]

14.8 Show that if X and Y are i.i.d. then $Z = X - Y$ has a symmetric distribution.

14.9 Let $X_1, \ldots, X_n$ be independent, each with mean 0, and each with finite third moments. Show that

$$E\left\{\left(\sum_{i=1}^n X_i\right)^3\right\} = \sum_{i=1}^n E\{X_i^3\}.$$

(*Hint:* Use characteristic functions.)

14.10 Let $\mu_1, \ldots, \mu_n$ be probability measures. Suppose $\lambda_j \geq 0$ $(1 \leq j \leq n)$ and $\sum_{j=1}^n \lambda_j = 1$. Let $\nu = \sum_{j=1}^n \lambda_j \mu_j$. Show that ν is a probability measure, too, and that

$$\hat{\nu}(u) = \sum_{j=1}^n \lambda_j \hat{\mu}_j(u).$$

14.11 Let X have the double exponential (or Laplace) distribution with $\alpha = 0$, $\beta = 1$:

$$f_X(x) = \frac{1}{2} e^{-|x|} \qquad -\infty < x < \infty.$$

Show that $\varphi_X(u) = \frac{1}{1+u^2}$.
   (*Hint:* Use Exercise 14.10 with μ_1 the distribution of Y and μ_2 the distribution of $-Y$, where Y is an Exponential of parameter $\lambda = 1$. (Take $\lambda_1 = \lambda_2 = \frac{1}{2}$.))

14.12 * (*Triangular distribution*) Let X be a r.v. with density $f_X(x) = (1 - |x|)1_{(-1,1)}(x)$. Show that $\varphi_X(u) = \frac{2(1-\cos u)}{u^2}$. (*Hint:* Let U, V be independent uniform on $(-\frac{1}{2}, \frac{1}{2})$ and consider $U + V$. Observe further that $\left(\frac{e^{iu/2} - e^{-iu/2}}{iu}\right)^2 = \left(\frac{2\sin(u/2)}{u}\right)^2$.)

14.13 Let X be a positive random variable. The *Mellin transform* of X is defined to be

$$T_X(\theta) = E\{X^\theta\}$$

for all values of θ for which the expected value of X^θ exists.

a) Show that

$$T_X(\theta) = \varphi_{\log X}\left(\frac{\theta}{i}\right)$$

   when all terms are well defined.
b) Show that if X and Y are independent and positive, then

$$T_{XY}(\theta) = T_X(\theta)T_Y(\theta).$$

c) Show that $T_{bX^a}(\theta) = b^\theta T_X(a\theta)$ for $b > 0$ and $a\theta$ in the domain of definition of T_X.

14.14 Let X be lognormal with parameters (μ, σ^2). Find the Mellin transform (c.f. Exercise 14.13) $T_X(\theta)$. Use this and the observation that $T_X(k) = E\{X^k\}$ to calculate the kth moments of the lognormal distribution for $k = 1, 2, \ldots$.

14.15 Let X be $N(0, 1)$. Show that $E\{X^{2n+1}\} = 0$ and

$$E\{X^{2n}\} = \frac{(2n)!}{2^n n!} = (2n - 1)(2n - 3) \ldots 3 \cdot 1.$$

14.16 * Let X be $N(0, 1)$. Let

$$M(s) = E\{e^{sX}\} = \int_{-\infty}^{\infty} \frac{1}{\sqrt{2\pi}} \exp\left(sx - \frac{1}{2}x^2\right) dx.$$

Show that $M(s) = e^{s^2/2}$. (*Hint:* Complete the square in the integrand.)

14.17 * Substitute $s = iu$ in Exercise 14.16 to obtain the characteristic function of the Normal $\varphi_X(u) = e^{-u^2/2}$; justify that one can do this by the theory of analytic continuation of functions of a complex variable.

14.18 Let X be $N(\mu, \sigma^2)$. Show that $Y = \frac{X - \mu}{\sigma}$ is $N(0, 1)$.

14.19 * (Feller [9]) Let X be a Gamma r.v. with parameters (α, β). One can calculate its characteristic function without using contour integration. Assume $\beta = 1$ and expand e^{ix} in a power series. Then show

$$\frac{1}{\Gamma(\alpha)} \sum_{n=1}^{\infty} \frac{(iu)^n}{n!} \int_0^{\infty} e^{-x} x^{n+\alpha-1} dx = \sum_{n=0}^{\infty} \frac{\Gamma(n + \alpha)}{n! \Gamma(\alpha)} (iu)^n$$

and show this is a binomial series which sums to $\frac{1}{(1-iu)^\alpha}$.

15 Sums of Independent Random Variables

Many of the important uses of Probability Theory flow from the study of sums of independent random variables. A simple example is from Statistics: if we perform an experiment repeatedly and independently, then the "average value" is given by $\bar{x} = \frac{1}{n}\sum_{j=1}^{n} X_j$, where X_j represents the outcome of the jth experiment. The r.v. $\bar{x}$ is then called an *estimator* for the mean μ of each of the X_j. Statistical theory studies when (and how) $\bar{x}$ converges to μ as n tends to ∞. Even once we show that $\bar{x}$ tends to μ as n tends to ∞, we also need to know how large n should be in order to be reasonably sure that $\bar{x}$ is close to the true value μ (which is, in general, unknown). There are other, more sophisticated questions that arise as well: what is the probability distribution of $\bar{x}$? If we cannot infer the exact distribution of $\bar{x}$, can we approximate it? How large need n be so that our approximation is sufficiently accurate? If we have prior information about μ, how do we use that to improve upon our estimator $\bar{x}$? Even to begin to answer some of these fundamentally important questions we need to study sums of independent random variables.

Theorem 15.1. *Let X, Y be two $\mathbf{R}$-valued independent random variables. The distribution measure μ_Z of $Z = X + Y$ is the convolution product of the probability measures μ_X and μ_Y, defined by*

$$\mu_X * \mu_Y(A) = \int\int 1_A(x + y)\mu_X(dx)\mu_Y(dy). \tag{15.1}$$

Proof. Since X and Y are independent, we know that the joint distribution of (X, Y) is $\mu_X \otimes \mu_Y$. Therefore

$$E\{g(X, Y)\} = \int\int g(x, y)\mu_X(dx)\mu_Y(dy),$$

and in particular, using $g(x, y) = f(x + y)$:

$$E\{f(X + Y)\} = \int\int f(x + y)\mu_X(dx)\mu_Y(dy), \tag{15.2}$$

for any Borel function f on $\mathbf{R}$ for which the integrals exist. It suffices to take $f(x) = 1_A(x)$. $\qquad\square$

Remark 15.1. Formula (15.2) above shows that for $f : \mathbf{R} \to \mathbf{R}$ Borel measurable and $Z = X + Y$ with X and Y independent:

$$E\{f(Z)\} = \int f(z)(\mu_X * \mu_Y)(dz) = \int\int f(x + y)\mu_X(dx)\mu_Y(dy).$$

Theorem 15.2. *Let X, Y be independent real valued random variables, with $Z = X + Y$. Then the characteristic function φ_Z is the product of φ_X and φ_Y; that is:*

$$\varphi_Z(u) = \varphi_X(u)\varphi_Y(u).$$

Proof. Let $f(z) = e^{i\langle u,z \rangle}$ and use formula (15.2).   □

Caution: If $Z = X + Y$, the property that $\varphi_Z(u) = \varphi_X(u)\varphi_Y(u)$ for all $u \in \mathbf{R}$ is not enough to ensure that X and Y are independent.

Theorem 15.3. *Let X, Y be independent real valued random variables and let $Z = X + Y$.*

a) *If X has a density f_X, then Z has a density f_Z and moreover:*

$$f_Z(z) = \int f_X(z - y)\mu_Y(dy)$$

b) *If in addition Y has a density f_Y, then*

$$f_Z(z) = \int f_X(z - y)f_Y(y)dy = \int f_X(x)f_Y(z - x)dx.$$

Proof. (b): Suppose (a) is true. Then

$$f_Z(z) = \int f_X(z - y)\mu_Y(dy).$$

However $\mu_Y(dy) = f_Y(y)dy$, and we have the first equality. Interchanging the roles of X and Y gives the second equality.

(a): By Theorem 15.1 we have

$$\mu_Z(A) = \int\int 1_A(x + y)\mu_X(dx)\mu_Y(dy)$$

$$= \int \left\{ \int 1_A(x + y)f_X(x)dx \right\} \mu_Y(dy).$$

Next let $z = x + y$; $dz = dx$;

$$= \int \left\{ \int 1_A(z)f_X(z - y)dz \right\} \mu_Y(dy)$$

and applying the Tonelli-Fubini theorem:

$$= \int \left\{ \int f(z - y)\mu_Y(dy) \right\} 1_A(z)dz.$$

Since A was arbitrary we have the result for all Borel sets A, which proves the theorem.   □

The next theorem is trivial but surprisingly useful.

Theorem 15.4. *Let X, Y be independent real valued random variables that are square integrable (that is $E\{X^2\} < \infty$ and $E\{Y^2\} < \infty$). Then*

$$\sigma^2_{X+Y} = \sigma^2_X + \sigma^2_Y.$$

Proof. Since X and Y are independent we have $E\{XY\} = E\{X\}E\{Y\}$, and

$$\sigma^2_{X+Y} = E\{X^2\} + 2E\{XY\} + E\{Y^2\} - (E\{X\} + E\{Y\})^2 = \sigma^2_X + \sigma^2_Y.$$

$\square$

Examples:

1. Let $X_1, \ldots, X_n$ be i.i.d. Bernoulli (p). Then $Y = \sum_{j=1}^n X_j$ is Binomial $B(p, n)$. We have seen

$$E\{Y\} = E\left\{\sum_{j=1}^n X_j\right\} = \sum_{j=1}^n E\{X_j\} = \sum_{j=1}^n p = np.$$

   Note that

$$\sigma^2_{X_j} = E\{X_j^2\} - E\{X_j\}^2 = p - p^2 = p(1 - p).$$

   Therefore by Theorem 15.4,

$$\sigma^2_Y = \sum_{j=1}^n \sigma^2_{X_j} = np(1 - p).$$

   Note that the above method of computing the variance is preferable to explicit use of the distribution of Y, which would give rise to the following calculation:

$$\sigma^2_Y = \sum_{j=0}^n (j - np)^2 \binom{n}{j} p^j (1 - p)^{n-j}.$$

2. Let X be Poisson (λ) and Y be Poisson (μ), and X and Y are independent. Then $Z = X + Y$ is also Poisson $(\lambda + \mu)$. Indeed, $\varphi_Z = \varphi_X \varphi_Y$ implies

$$\varphi_Z(u) = e^{\lambda(e^{iu} - 1)} e^{\mu(e^{iu} - 1)} = e^{(\lambda + \mu)(e^{iu} - 1)},$$

   which is the characteristic function of a Poisson $(\lambda + \mu)$. Therefore Z is Poisson by the uniqueness of characteristic functions (Theorem 14.1).

3. Suppose X is Binomial $B(p, n)$ and Y is Binomial $B(p, m)$. (X and Y have the same p.) Let $Z = X + Y$. Then

$$\varphi_Z = \varphi_X \varphi_Y,$$

   hence

$$\varphi_Z(u) = (pe^{iu} + (1-p))^n(pe^{iu} + (1-p))^m = (pe^{iu} + (1-p))^{n+m},$$

which is the characteristic function of a Binomial $B(p, m+n)$; hence Z is Binomial $B(p, m+n)$ by Theorem 14.1. We did not really need characteristic functions for this result: simply note that

$$X = \sum_{j=1}^{n} U_j \qquad \text{and} \qquad Y = \sum_{j=1}^{m} V_j,$$

and thus

$$Z = \sum_{j=1}^{n} U_j + \sum_{j=1}^{m} V_j,$$

where U_j and V_j are all i.i.d. Bernoulli (p). Hence

$$Z = \sum_{j=1}^{m+n} W_j$$

where W_j are i.i.d. Bernoulli (p). (The first n W_j's are the U_j's; the next m W_j's are the V_j's.)

4. Suppose X is normal $N(\mu, \sigma^2)$ and Y is also normal $N(\nu, \tau^2)$, and X and Y are independent. Then $Z = X + Y$ is normal $N(\mu+\nu, \sigma^2+\tau^2)$. Indeed

$$\varphi_Z = \varphi_X\varphi_X$$

implies

$$\varphi_Z(u) = e^{iu\mu - u^2\sigma^2/2}e^{iu\nu - u^2\tau^2/2} = e^{iu(\mu+\nu) - u^2(\sigma^2+\tau^2)/2}$$

which is the characteristic function of a normal $N(\mu + \nu, \sigma^2 + \tau^2)$, and we again use Theorem 14.1.

5. Let X be the Gamma (α, β) and Y be Gamma (δ, β) and suppose X and Y are independent. Then if $Z = X + Y$, $\varphi_Z = \varphi_X\varphi_Y$, and therefore

$$\varphi_Z(u) = \frac{\beta^\alpha}{(\beta - iu)^\alpha}\frac{\beta^\delta}{(\beta - iu)^\delta} = \frac{\beta^{\alpha+\delta}}{(\beta - iu)^{\alpha+\delta}},$$

whence Z has the characteristic function of a Gamma $(\alpha+\delta, \beta)$, and thus by Theorem 14.1, Z is a Gamma $(\alpha + \delta, \beta)$.

6. In Chapter 11 we defined the *chi square distribution with p degrees of freedom* (denoted χ_p^2), and we observed that if X is χ_1^2, then $X = Z^2$ in distribution, where Z is $N(0, 1)$. We also noted that if X is χ_p^2, then X is Gamma $(\frac{p}{2}, \frac{1}{2})$. Therefore let X be χ_p^2, and let $Z_1, \ldots, Z_p$ be i.i.d. $N(0, 1)$. If $Y = \sum_{i=1}^{p} Z_i^2$, by Example 5 we have that since each Z_i^2 is Gamma $(\frac{1}{2}, \frac{1}{2})$, then Y is Gamma $(\frac{p}{2}, \frac{1}{2})$ which is χ_p^2. We conclude that if X is χ_p^2, then $X = \sum_{i=1}^{p} Z_i^2$ in distribution, where Z_i are i.i.d. $N(0, 1)$.

Exercises for Chapter 15

15.1 Let $X_1, \ldots, X_n$ be independent random variables, and assume $E\{X_j\} = \mu$ and $\sigma^2(X_j) = \sigma^2 < \infty$, $1 \leq j \leq n$. Let

$$\bar{x} = \frac{1}{n} \sum_{j=1}^n X_j \qquad \text{and} \qquad S^2 = \frac{1}{n} \sum_{j=1}^n (X_j - \bar{x})^2.$$

($\bar{x}$ and S^2 are also random variables, known as the "sample mean" and the "sample variance", respectively.) Show that

a) $E\{\bar{x}\} = \mu$;
b) $\mathrm{Var}\,(\bar{x}) = \frac{\sigma^2}{n}$;
c) $E(S^2) = \frac{n-1}{n}\sigma^2$.

15.2 Let $X_1, \ldots, X_n$ be independent with finite variances. Let $S_n = \sum_{j=1}^n X_j$. Show that

$$\sigma^2_{\frac{1}{n} S_n} = \frac{1}{n^2} \sum_{j=1}^n \sigma^2_{X_j},$$

and deduce that if $\sigma^2_{X_j} = \sigma^2$, $1 \leq j \leq n$, then $\sigma^2_{\frac{1}{n} S_n} = \sigma^2/n$.

15.3 Show that if $X_1, \ldots, X_n$ are i.i.d., then

$$\varphi_{S_n}(u) = (\varphi_X(u))^n,$$

where $S_n = \sum_{j=1}^n X_j$.

Problems 4–8 involve the summation of a random number of independent random variables. We let $X_1, X_2, \ldots$ be an infinite sequence of i.i.d. random variables and let N be a positive, integer-valued random variable which is independent from the sequence. Further, let

$$S_n = \sum_{i=1}^n X_i, \qquad \text{and} \qquad S_N = X_1 + X_2 + \ldots + X_N,$$

with the convention that $S_N = 0$ if $N = 0$.

15.4 For a Borel set A, show that

$$P(S_N \in A \mid N = n) = P(S_n \in A).$$

15.5 Suppose $E\{N\} < \infty$ and $E\{|X_j|\} < \infty$. Show

$$E\{S_N\} = \sum_{n=0}^{\infty} E\{S_n\}P(N = n).$$

(*Hint:* Show first that

$$E\{S_N\} = E\left\{\sum_{n=0}^{\infty} S_n 1_{\{N=n\}}\right\} = \sum_{n=0}^{\infty} E\{S_n 1_{\{N=n\}}\},$$

and justify the second equality above.)

15.6 Suppose $E\{N\} < \infty$ and $E\{|X_j|\} < \infty$. Show that $E\{S_N\} = E\{N\}E\{X_j\}$. (*Hint:* Use Exercise 15.5.)

15.7 Suppose $E\{N\} < \infty$ and $E\{|X_j|\} < \infty$. Show that

$$\varphi_{S_N}(u) = E\{(\varphi_{X_j}(u))^N\}.$$

(*Hint:* Show first

$$\varphi_{S_N}(u) = \sum_{n=1}^{\infty} E\{e^{iuS_n} 1_{\{N=n\}}\}.)$$

15.8 Solve Exercise 15.6 using Exercise 15.7. (*Hint:* Recall that $E\{Z\} = i\varphi_Z'(0)$, for a r.v. Z in L^1.)

15.9 Let X, Y be real valued and independent. Suppose X and $X + Y$ have the same distribution. Show that Y is a constant r.v. equal to 0 a.s.

15.10 Let f, g map $\mathbf{R}$ to $\mathbf{R}_+$ such that

$$\int_{-\infty}^{\infty} f(x)dx < \infty \quad \text{and} \quad \int_{-\infty}^{\infty} g(x)dx < \infty.$$

Show that

a) $f * g(x) = \int_{-\infty}^{\infty} f(x - y)g(y)dy$ exists;
b) $f * g(x) = g * f(x)$ ($f * g$ is called the *convolution* of f and g)
c) If one of f or g is continuous, then $f * g$ is continuous.

15.11 Let X, Y be i.i.d. Suppose further that $X + Y$ and $X - Y$ are independent. Show that $\varphi_X(2u) = (\varphi_X(u))^3 \varphi_X(-u)$.

15.12 * Let X, Y be as in Exercise 15.11, and also that $E\{X\} = 0$ and $E\{X^2\} = 1$. Show that X is Normal $N(0, 1)$. (*Hint:* Show that for some $a > 0$ we have $\varphi(u) \neq 0$ for all u with $|u| \leq a$. Let $\psi(u) = \frac{\varphi(u)}{\varphi(-u)}$ for $|u| \leq a$, and show $\psi(u) = \{\psi(u/2^n)\}^{2^n}$; then show this tends to 1 as $n \to \infty$. (See Exercise 14.4.) Deduce that $\varphi(t) = \{\varphi(t/2^n)\}^{4^n}$ and let $n \to \infty$.)

15.13 Let $X_1, X_2, \ldots, X_n$ be i.i.d. Normal $N(\mu, \sigma^2)$. Let $\bar{x} = \frac{1}{n}\sum_{j=1}^{n} X_j$ and let $Y_j = X_j - \bar{x}$. Find the joint characteristic function of $(\bar{x}, Y_1, \ldots, Y_n)$. Let $S^2 = \frac{1}{n}\sum_{j=1}^{n} Y_j^2$. Deduce that $\bar{x}$ and S^2 are independent.

15.14 Show that $|1 - e^{ix}|^2 = 2(1 - \cos x) \le x^2$ for all $x \in \mathbf{R}$. Use this to show that $|1 - \varphi_X(u)| \le E\{|uX|\}$.

15.15 Let $A = [-\frac{1}{u}, \frac{1}{u}]$. Show that

$$\int_A x^2 \mu_X(dx) \le \frac{12}{11u^2}\{1 - \mathrm{Re}\ \varphi_X(u)\}.$$

(*Hint:* $1 - \cos x \ge 0$ and $1 - \cos x \ge \frac{1}{2}x^2 - \frac{1}{24}x^4$, all $x \in \mathbf{R}$; also if $z = a + ib$, then Re $z = a$, where $a, b, \in \mathbf{R}$.)

15.16 If φ is a characteristic function, show that $|\varphi|^2$ is one too. (*Hint:* Let X, Y be i.i.d. and consider $Z = X - Y$.)

15.17 Let $X_1, \ldots, X_\alpha$ be independent exponential random variables with parameter $\beta > 0$. Show that $Y = \sum_{i=1}^{\alpha} X_i$ is Gamma (α, β).

16 Gaussian Random Variables (The Normal and the Multivariate Normal Distributions)

Let us recall that a Normal random variable with parameters (μ, σ^2), where $\mu \in \mathbf{R}$ and $\sigma^2 > 0$, is a random variable whose density is given by:

$$f(x) = \frac{1}{\sqrt{2\pi}\sigma} e^{-(x-\mu)^2/2\sigma^2}, \quad -\infty < x < \infty. \tag{16.1}$$

Such a distribution is usually denoted $N(\mu, \sigma^2)$. For convenience of notation, we extend the class of normal distributions to include the parameters $\mu \in \mathbf{R}$ and $\sigma^2 = 0$ as follows: we will denote by $N(\mu, 0)$ the law of the constant r.v. equal to μ (this is also the Dirac measure at point μ). Of course, the distribution $N(\mu, 0)$ has no density, and in this case we sometimes speak of a *degenerate normal distribution*. When $\mu = 0$ and $\sigma^2 = 1$, we say that $N(0, 1)$ is the *standard Normal distribution*.

When X is a r.v. with distribution $N(\mu, \sigma^2)$ we write $X \overset{\mathcal{D}}{=} N(\mu, \sigma^2)$, or alternatively $\mathcal{L}(X) = N(\mu, \sigma^2)$, where $\mathcal{L}$ stands for "law": that is, "the law of X is $N(\mu, \sigma^2)$". The characteristic function φ_X of X is

$$\varphi_X(u) = e^{iu\mu - \frac{\sigma^2 u^2}{2}}. \tag{16.2}$$

When $\sigma^2 > 0$ this comes from Example 13.6, and when $\sigma^2 = 0$ this is trivial. Let us recall also that when $\mathcal{L}(X) = N(\mu, \sigma^2)$, then

$$E\{X\} = \mu, \quad \mathrm{Var}(X) = \sigma^2. \tag{16.3}$$

At first glance it might seem strange to call a distribution with such an odd appearing density "normal". The reason for this dates back to the early 18th century, when the first versions of the Central Limit Theorem appeared in books by Jacob Bernoulli (1713) and A. de Moivre (1718). These early versions of the Central Limit Theorem were expanded upon by P. Laplace and especially C. F. Gauss. Indeed because of the fundamental work of Gauss normal random variables are often called *Gaussian random variables,* and the former 10 Deutsche Mark note in Germany has a picture of Gauss on it and a graph of the function f given in (16.1), which is known as the *Gaussian density.* (This use of Probability Theory on currency, perhaps inspired by the extensive use of probability in Finance, has disappeared now that the Mark has been replaced with the Euro.) The Gaussian version of the Central Limit

Theorem can be loosely interpreted as saying that sums of i.i.d. random variables are approximately Gaussian. This is quite profound, since one needs to know almost nothing about the actual distributions one is summing to conclude the sum is approximately Gaussian. Finally we note that later Paul Lévy did much work to find minimal hypotheses for the Central Limit theorem to hold. It is this family of theorems that is central to much of Statistical Theory; it allows one to assume a precise Gaussian structure from minimal hypotheses. It is the "central" nature of this theorem in Statistics that gives it its name, and which in turn makes Gaussian random variables both important and ubiquitous, hence normal. We treat the Central Limit Theorem in Chapter 21; here we lay the groundwork by studying the Gaussian random variables that will arise as the limiting distributions.

For a real-valued random variable X the definition $\mathcal{L}(X) = N(\mu, \sigma^2)$ is clear: it is a r.v. X with a density given by (16.1) if $\sigma^2 > 0$, and it is $X \equiv \mu$ if $\sigma^2 = 0$. For an $\mathbf{R}^n$-valued r.v. the definition is more subtle; the reason is that we are actually describing the class of random variables that can arise as limits in the Central Limit Theorem, and this class is more complicated in $\mathbf{R}^n$ when $n \geq 2$.

Definition 16.1. *An $\mathbf{R}^n$-valued random variable $X = (X_1, \ldots, X_n)$ is Gaussian (or Multivariate Normal) if every linear combination $\sum_{j=1}^{n} a_j X_j$ has a (one-dimensional) Normal distribution (possibly degenerate; for example it has the distribution $N(0, 0)$ when $a_j = 0$, all j).*

Characteristic functions are of help when studying Gaussian random variables.

Theorem 16.1. *X is an $\mathbf{R}^n$-valued Gaussian random variable if and only if its characteristic function has the form*

$$\varphi_X(u) = \exp\{i\langle u, \mu \rangle - \frac{1}{2}\langle u, Qu \rangle\} \tag{16.4}$$

where $\mu \in \mathbf{R}^n$ and Q is an $n \times n$ symmetric nonnegative semi-definite matrix. Q is then the covariance matrix of X and μ is the mean of X, that is $\mu_j = E\{X_j\}$ for all j.

Proof (Sufficiency): Suppose we have (16.4). Let

$$Y = \sum_{j=1}^{n} a_j X_j = \langle a, X \rangle$$

be a linear combination of the components of X. We need to show Y is (univariate) normal. But then for $v \in \mathbf{R}$:

$$\varphi_Y(v) = \varphi_X(va) = \exp\left\{iv\langle a, \mu \rangle - \frac{v^2}{2}\langle a, Qa \rangle\right\}$$

and by equation (16.2), $\varphi_Y(v)$ is the characteristic function of a normal $N(\langle a, \mu \rangle, \langle a, Qa \rangle)$, and thus by Theorem 14.1 we have Y is normal.

(Necessity): Suppose X is Gaussian, and let

$$Y = \sum_{j=1}^{n} a_j X_j = \langle a, X \rangle$$

be a linear combination of the components of X. Let $Q = \mathrm{Cov}(X)$ be the covariance matrix of X. Then

$$E\{Y\} = \langle a, \mu \rangle$$

where $\mu = (\mu_1, \ldots, \mu_n)$ and $E\{X_i\} = \mu_i$, $1 \le i \le n$; and also

$$\sigma^2(Y) = \langle a, Qa \rangle,$$

by Theorem 12.4. Since Y is normal by hypothesis, by (16.2) again we have

$$\varphi_Y(v) = \exp\left\{ iv\langle a, \mu \rangle - \frac{v^2}{2} \langle a, Qa \rangle \right\}.$$

Then

$$\varphi_Y(1) = \varphi_{\langle a, X \rangle}(1) = E\{\exp(i\langle a, X \rangle)\} = \varphi_X(a),$$

and we have equation (16.4). $\square$

Notation: When X is as in the previous theorem, we denote by $N(\mu, Q)$ its law. Its depends on the two parameters: μ, the *mean vector*, and Q, the *covariance matrix*. The terminology "mean vector" is clear from the previous proof, in which we have seen that $\mu_i = E\{X_i\}$. For the matrix $Q = (Q_{i,j})_{1 \le i,j \le n}$, by differentiating Equation (16.4) twice and by using Theorem 13.2 we obtain

$$E\{X_i X_j\} = \mu_i \mu_j + Q_{i,j}.$$

Then we see that $Q_{i,j} = \mathrm{Cov}(X_i, X_j)$.

Example 1:

Let $X_1, \ldots, X_n$ be $\mathbf{R}$-valued *independent* random variables with laws $N(\mu_j, \sigma_j^2)$. Then $X = (X_1, \ldots, X_n)$ is Gaussian (i.e., Multivariate Normal). This is easy to verify, since

$$\varphi_X(u_1, \ldots, u_n) = \prod_{j=1}^{n} \varphi_{X_j}(u_j)$$

by Corollary 14.1; therefore

$$\varphi_X(u) = \prod_{j=1}^n e^{iu_j\mu_j - u_j^2\sigma_j^2/2}$$

$$= \exp\left(\sum_{j=1}^n iu_j\mu_j - \frac{1}{2}\sum_{j=1}^n u_j^2\sigma_j^2\right)$$

$$= e^{i\langle u,\mu\rangle - \frac{1}{2}\langle u, Qu\rangle}$$

where $\mu = (\mu_1, \ldots, \mu_n)$ and Q is the diagonal matrix

$$\begin{pmatrix} \sigma_1^2 & & & 0 \\ & \sigma_2^2 & & \\ & & \ddots & \\ 0 & & & \sigma_n^2 \end{pmatrix}.$$

Since $\varphi_X(u)$ is of the form (16.4), we know that X is multivariate normal.

The converse of Example 1 is also true:

Corollary 16.1. *Let X be an $\mathbf{R}^n$-valued Gaussian random variable. The components X_j are independent if and only if the covariance matrix Q of X is diagonal.*

Proof. Example 1 shows the necessity. Suppose then we know Q is diagonal, i.e.

$$Q = \begin{pmatrix} \sigma_1^2 & & & 0 \\ & \sigma_2^2 & & \\ & & \ddots & \\ 0 & & & \sigma_n^2 \end{pmatrix}.$$

By Equation (16.4) of Theorem 16.1 it follows that φ_X factors:

$$\varphi_X(u) = \prod_{j=1}^n \varphi_{X_j}(u_j),$$

where

$$\varphi_{X_j}(u_j) = \exp\left\{iu_j\mu_j - \frac{1}{2}u_j^2\sigma_j^2\right\}.$$

Corollary 14.1 then gives that the X_j are independent, and they are each normal ($N(\mu_j, \sigma_j^2)$) by Equation (16.2). $\qquad\square$

The next theorem shows that *all* Gaussian random variables (i.e., Multivariate Normal random variables) arise as linear transformations of vectors of independent univariate normals. (Recall that we use the terms normal and Gaussian interchangeably.)

Theorem 16.2. *Let X be an $\mathbf{R}^n$-valued Gaussian random variable with mean vector μ. Then there exist independent Normal random variables Y_1, ..., Y_n with*

$$\mathcal{L}(Y_j) = N(0, \lambda_j), \quad \lambda_j \geq 0, \quad (1 \leq j \leq n),$$

and an orthogonal matrix A such that $X = \mu + AY$.

Important Comment: We have assumed in Theorem 16.2 only that $\lambda_j \geq 0$. Some of the λ_j can sometimes take the value zero. In this case we have $Y_j = 0$. Thus the number of independent normal random variables required in Theorem 16.2 can be strictly less in number than the number of non-trivial components in the Gaussian r.v. X.

Proof of Theorem 16.2: Since Q is a covariance matrix it is symmetric, non-negative semi-definite and there always exists an orthogonal matrix A such that $Q = A\Lambda A^*$, where Λ is a diagonal matrix with all entries nonnegative. (Recall that an *orthogonal matrix* is a matrix where the rows (or columns), considered as vectors, are orthonormal: that is they have length (or norm) one and the scalar product of any two of them is zero (i.e., they are orthogonal).) A^* means the transpose of the matrix A. Since A is orthogonal, then A^* is also the inverse of A.

We set

$$Y = A^*(X - \mu)$$

where $\mu_j = E\{X_j\}$ for X_j the jth component of X. Since X is Gaussian by hypothesis, we have that Y must be Gaussian too, since any linear combination of the components of Y is also a linear combination of the components of X and therefore univariate normal. Moreover the covariance matrix of Y is $A^*QA = \Lambda$, the sought after diagonal matrix. Since $X = \mu + AY$ (because $A^{*-1} = A$), we have proved the theorem. $\qquad \square$

Corollary 16.2. *An $\mathbf{R}^n$-valued Gaussian random variable X has a density on $\mathbf{R}^n$ if and only if the covariance matrix Q is non-degenerate (that is, there does not exist a vector $a \in \mathbf{R}^n$ such that $Qa = 0$, or equivalently that $\det(Q) \neq 0$).*

Proof. By Theorem 16.2 we know there exist n independent normals $Y_1, \ldots, Y_n$ of laws $N(0, \lambda_j)$, $(1 \leq j \leq n)$, with $Q = A\Lambda A^*$, for an orthogonal matrix A. If $\det(Q) \neq 0$, we must have $\lambda_j > 0$, for all j $(1 \leq j \leq n)$, because $\det(Q) = \det(\Lambda) = \prod_{i=1}^n \lambda_i$. Since $\lambda_j > 0$ and $\mathcal{L}(Y_j) = N(0, \lambda_j)$, we know that Y has a density given by

$$f_Y(y) = \prod_{j=1}^n \frac{1}{\sqrt{2\pi\lambda_j}} e^{-y_j^2/2\lambda_j},$$

and since $X = \mu + AY$, we deduce from Theorem 12.7 that X has the density

$$f_X(x) = \frac{1}{2\pi^{n/2}\sqrt{\det Q}} e^{-\frac{1}{2}\langle x-\mu, Q^{-1}(x-\mu)\rangle}. \tag{16.5}$$

Next suppose Q is degenerate: that is, $\det(Q) = 0$. Then there exists an $a \in \mathbf{R}^n$, $a \neq 0$ such that $Qa = 0$ (that is, the kernel of the linear transformation represented by Q is non-trivial). The random variable $Z = \langle a, X \rangle$ has a variance equal to $\langle a, Qa \rangle = 0$, so it is a.s. equal to its mean $\langle a, \mu \rangle$. Therefore $P(X \in H) = 1$, where H is an affine hyperplane orthogonal to the vector a and containing the vector μ, that is $H = \{x \in \mathbf{R}^n : \langle x - \mu, a \rangle = 0\}$.) Since the dimension of H is $n - 1$, the n-dimensional Lebesgue measure of H is zero. If X were to have a density, we would need to have the property

$$1 = P(X \in H) = \int_H f(x)dx = \int_H f(x_1, \dots, x_n)dx_1 \dots dx_n. \tag{16.6}$$

However

$$\int 1_H(x_1, \dots, x_n)dx_1 \dots dx_n = 0$$

because H is a hyperplane (see Exercise 16.1), hence (16.6) cannot hold; whence X cannot have a density.   □

Comment: Corollary 16.2 shows that when $n \geq 2$ there exist normal (Gaussian) non constant random variables without densities (when $n = 1$ a normal variable is either constant or with a density). Moreover since (as we shall see in Chapter 21) these random variables arise as limits in the Central Limit Theorem, they are important and cannot be ignored. Thus while it is tempting to define Gaussian random variables as (for example) random variables having densities of the form given in (16.5), such a definition would not cover some important cases.

An elementary but important property of $\mathbf{R}^n$-valued Gaussian random variables is as follows:

Theorem 16.3. *Let X be an $\mathbf{R}^n$-valued Gaussian random variable, and let Y be an $\mathbf{R}^m$-valued Gaussian r.v. If X and Y are independent then $Z = (X, Y)$ is an $\mathbf{R}^{n+m}$-valued Gaussian r.v.*

Proof. We have

$$\varphi_Z(u) = \varphi_X(w)\varphi_Y(v)$$

where

$$u = (w, v); \quad w \in \mathbf{R}^n, \quad v \in \mathbf{R}^m;$$

since X and Y are independent. By Theorem 16.1

$$\varphi_Z(u) = \exp\left\{i\langle w, \mu^X\rangle - \frac{1}{2}\langle w, Q^X w\rangle\right\} \exp\left\{i\langle v, \mu^Y\rangle - \frac{1}{2}\langle v, Q^Y v\rangle\right\}$$

$$= \exp\left\{i\langle (w, v), (\mu^X, \mu^Y)\rangle - \frac{1}{2}\langle u, Qu\rangle\right\},$$

where
$$Q = \begin{pmatrix} Q^X & 0 \\ 0 & Q^Y \end{pmatrix}.$$

Again using Theorem 16.1 this is the characteristic function of a Gaussian r.v.   □

We say that two random variables X, Y are *uncorrelated* if Cov $(X, Y) = 0$. Since
$$\text{Cov}(X, Y) = E\{XY\} - E\{X\}E\{Y\},$$

this is equivalent to saying that $E\{XY\} = E\{X\}E\{Y\}$. This of course is true if X and Y are independent (Theorem 12.3) and thus X, Y *independent implies that X, Y are uncorrelated.* The converse is *false* is general. However it is true for the Multivariate Normal (or Gaussian) case, a surprising and useful fact.

Theorem 16.4. *Let X be an $\mathbf{R}^n$-valued Gaussian random variable. Two components X_j and X_k of X are independent if and only if they are uncorrelated.*

Proof. We have already seen the necessity. Conversely, suppose that X_j and X_k are uncorrelated. We can consider the two dimensional random vector $Y = (Y_1, Y_2)$, with $Y_1 = X_j$ and $Y_2 = X_k$. Clearly Y is a bivariate normal vector, and since $\text{Cov}(Y_1, Y_2) = 0$ by hypothesis we have that the covariance matrix of Y is diagonal, and the theorem reduces to Corollary 16.1.   □

A standard model used in science, engineering, and the social sciences is that of *simple linear regression*.[1] Here one has random variables $Y_i, 1 \leq i \leq n$, of the form
$$Y_i = \alpha + \beta x_i + \varepsilon_i, \qquad (1 \leq i \leq n) \tag{16.7}$$

where α, β, and x_i are all constants, and ε_i are random variables. A typical model is to think that one is measuring $\alpha + \beta x_i$ and one makes a measurement error ε_i. Because of the Central Limit Theorem (see Chapter 21), one often assumes that $(\varepsilon_i)_{1 \leq i \leq n}$ have a multivariate normal distribution. Following Berger and Casella [5] we call x_i the *predictor variable* (again, x_i is non-random), and we call Y_i the response variable.[2]

Let us assume that $E\{\varepsilon_i\} = 0, 1 \leq i \leq n$. Then taking expectations in (16.7) we have
$$E\{Y_i\} = \alpha + \beta x_i. \tag{16.8}$$

Typically one wishes to learn the nature of the linear relation between Y_i and x_i; this would be obvious if ε_i were not present to obscure it. That is, one

---

[1] The "linearity" refers to the linear nature of the dependence on the parameters α and β; not of x_i.

[2] Sometimes x_i is called the "independent variable" and Y_i the "dependent variable". We do not use this terminology because of the possible confusion with independent random variables.

wants to find α and β by observing Y_i and knowing x_i, $1 \leq i \leq n$. And to begin with, we rule out the case where all x_i's are equal, because then we can at best find the constant $\alpha + \beta x_1$ and we cannot discriminate between α and β: in other words, we suppose that $\sum_{i=1}^{n}(x_i - \bar{x})^2 > 0$, where $\bar{x} = \frac{1}{n}\sum_{i=1}^{n} x_i$.

One can treat these models quite generally (see Chapter 12 of [5] for example), but we will limit our attention to the most important case: where the "errors" are normal.

Indeed, let the random variables $(\varepsilon_i)_{1 \leq i \leq n}$ be i.i.d. $N(0, \sigma^2)$ random variables. An *estimator* U for α or V for β is a random variable which depends on the observed values $Y_1, \ldots, Y_n$ and possibly on the known numbers $x_1, \ldots, x_n$, but *not* on the unknown α and β. Among all possible estimators, the simplest ones are the so-called *linear estimators*, which are of the form

$$U = u_0 + \sum_{i=1}^{n} u_i Y_i, \qquad V = v_0 + \sum_{i=1}^{n} v_i Y_i \qquad (16.9)$$

for some sequences of constants $u_0, \ldots, u_n$ and $v_0, \ldots, v_n$. The estimators U and V are said to be *unbiased* if $E\{U\} = \alpha$ and $E\{V\} = \beta$. Since $E\{\varepsilon_i\} = 0$, this is the case if and only if

$$\alpha = E(U) = u_0 + \alpha \left(\sum_{i=1}^{n} u_i\right) + \beta \left(\sum_{i=1}^{n} u_i x_i\right) \qquad (16.10)$$

$$\beta = E(V) = v_0 + \alpha \left(\sum_{i=1}^{n} v_i\right) + \beta \left(\sum_{i=1}^{n} v_i x_i\right). \qquad (16.11)$$

These equations should be satisfied for all choices of α and β, and this is the case if and only if

$$u_0 = 0 \quad \sum_{i=1}^{n} u_i = 1 \quad \text{and} \quad \sum_{i=1}^{n} u_i x_i = 0, \qquad (16.12)$$

$$v_0 = 0 \quad \sum_{i=1}^{n} v_i = 0 \quad \text{and} \quad \sum_{i=1}^{n} v_i x_i = 1. \qquad (16.13)$$

Among those estimators U (resp. V) of the form (16.9), which satisfy (16.12) (resp. (16.13)), how do we choose one? A standard method is to compare squared error loss: that is, U_1 and V_1 can be considered to be "better" than U_2 and V_2 if

$$E\{(U_1 - \alpha)^2\} \leq E\{(U_2 - \alpha)^2\}, \quad E\{(V_1 - \beta)^2\} \leq E\{(V_2 - \beta)^2\}. \quad (16.14)$$

It is a standard exercise in a statistics course to show that

$$v_i = \frac{(x_i - \bar{x})}{\sum_{j=1}^{n}(x_j - \bar{x})^2}, \qquad u_i = \frac{1}{n} - \bar{x}v_i \qquad (16.15)$$

minimize the squared error losses, subject to the unbiasedness conditions (16.12) and (16.13), and thus give the best estimators among linear unbiased estimators. See, e.g., [5, pp.557–564]. Because of (16.15) these best estimators take the following form, with $\overline{Y} = \frac{1}{n} \sum_{i=1}^{n} Y_i$:

$$B = \sum_{i=1}^{n} \frac{(x_i - \overline{x})}{\sum_{j=1}^{n}(x_j - \overline{x})^2} Y_i, \qquad A = \overline{Y} - \overline{x}B. \tag{16.16}$$

We can now determine the distributions of the estimates B and A for the Gaussian case. It is a spectacular property of the Gaussian that A and B are also themselves jointly Gaussian! Thus we can not only determine the distributions of our estimato

Theorem 16.5. *Let $(\varepsilon_i)_{1 \leq i \leq n}$ be i.i.d. $N(0, \sigma^2)$ and suppose*

$$Y_i = \alpha + \beta x_i + \varepsilon_i, \qquad (1 \leq i \leq n).$$

The estimators B for β and A for α given in (16.16) are jointly Gaussian, with means α and β respectively, and covariance matrix given by

$$\mathrm{Var}\,(A) = \frac{\sigma^2}{n \sum_{j=1}^{n}(x_j - \overline{x})^2} \sum_{j=1}^{n} x_j^2,$$

$$\mathrm{Var}\,(B) = \frac{\sigma^2}{\sum_{j=1}^{n}(x_j - \overline{x})^2},$$

$$\mathrm{Cov}\,(A, B) = \frac{-\sigma^2 \overline{x}}{\sum_{j=1}^{n}(x_j - \overline{x})^2},$$

where $\overline{x} = \frac{1}{n} \sum_{j=1}^{n} x_j$.

Proof. Since the $(\varepsilon_i)_{1 \leq i \leq n}$ are independent and Gaussian, they form a Gaussian vector by Example 1. Now, A and B are both equal to a constant plus a linear combination of the ε_i's, so the pair (A, B) is Gaussian by the very definition. By construction they are unbiased estimators for α and β, that is they have α and β for their respective means.

Since for any r.v. U we have $\mathrm{Var}\,(uU + v) = u^2 \mathrm{Var}\,(U)$, we can write

$$\mathrm{Var}\,(B) = \mathrm{Var}\left(\sum_{i=1}^{n} v_i Y_i\right)$$

$$= \mathrm{Var}\left(\sum_{i=1}^{n} v_i \varepsilon_i\right)$$

$$= \sum_{i=1}^{n} v_i^2 \mathrm{Var}\,(\varepsilon_i) \qquad \text{by independence of the } \varepsilon_i\text{'s}$$

$$= \sigma^2 \sum_{i=1}^{n} \left(\frac{(x_i - \overline{x})}{\sum_{j=1}^{n}(x_j - \overline{x})^2} \right)^2$$

$$= \frac{\sigma^2}{\sum_{i=1}^{n}(x_i - \overline{x})^2}.$$

Similarly, we have

$$\text{Var}\,(A) = \sum_{i=1}^{n} v_i^2 \text{Var}\,(\varepsilon_i)$$

$$= \sigma^2 \sum_{i=1}^{n} \left(\frac{1}{n^2} + \overline{x}^2 v_i^2 - 2\frac{\overline{x}v_i}{n} \right)$$

$$= \sigma^2 \left(\frac{1}{n} + \overline{x}^2 \sum_{i=1}^{n} v_i^2 \right)$$

$$= \frac{\sigma^2 \sum_{i=1}^{n} x_i^2}{\sum_{i=1}^{n}(x_i - \overline{x})^2}.$$

$$\text{Cov}\,(A, B) = \text{Cov}\,\left(\sum_{i=1}^{n} u_i \varepsilon_i, \sum_{i=1}^{n} v_i \varepsilon_i \right)$$

$$= \sum_{i=1}^{n} u_i v_i \text{Var}\,(\varepsilon_i)$$

$$= \sigma^2 \sum_{i=1}^{n} \left(\frac{v_i}{n} - \overline{x}v_i^2 \right) = -\sigma^2 \overline{x} \sum_{i=1}^{n} v_i^2 = -\overline{x}\text{Var}\,(B)$$

□

We end this chapter with an example that serves as a *warning*: X can be an $\mathbf{R}^n$-valued random variable, with each component being univariate normal, but *not be multivariate normal* (or Gaussian). *Thus the property of being multivariate normal is stronger than simply having each component being normal.*

Example 2. Let $\mathcal{L}(Y) = N(0, 1)$, and set for some $a > 0$:

$$Z = Y1_{\{|Y| \le a\}} - Y1_{\{|Y| > a\}}.$$

Then Z is also $N(0, 1)$ (see Exercise 16.2), but $Y + Z = 2Y1_{\{|Y| \le a\}}$ which is not normal, since (for example) $P(Y + Z > 2|a|) = 0$ and $Y + Z$ is not a.s. equal to a constant. Therefore $X = (Y, Z)$ is an $\mathbf{R}^2$-valued r.v. which is not Gaussian, even though its two components are each normal (or Gaussian).

It is worth emphasizing that the Multivariate Normal has several special properties not shared in general with other distributions. We have seen that

1. Components are independent if and only if they are uncorrelated;
2. We have that the components are univariate normal: thus the components belong to the same distribution family as the vector random variable;
3. A Gaussian X with an $N(\mu, Q)$ distribution with Q invertible can be linearly transformed into an $N(0, I)$ r.v. (Exercise 16.6); that is, linear transformations *do not* change the distribution family;
4. The density exists if and only if the covariance matrix is nondegenerate, giving a simple criterion for the existence of a density; and finally
5. We have that the conditional distributions of Multivariate Normal distributions are also normal (Exercise 16.10).

These six properties show a remarkable stability inherent in the Multivariate Normal. There are many more special features of the normal that we do not go into here.

It is interesting to note that the normal distribution does not really exist in nature. It arises via a limiting procedure (the Central Limit Theorem), and thus it is an approximation of reality, and often it is an excellent approximation. When one says, for example, that the heights of twenty year old men in the United States are normally distributed with mean μ and variance σ^2, one actually means that the heights are *approximately* so distributed. Indeed, if the heights were in fact normally distributed, there would be a strictly positive probability of finding men that were of negative height and also of finding men taller than the Sears Tower in Chicago. Such results are of course nonsense. However these positive probabilities are so small as to be equal to zero to many decimal places, and since zero is the true probability of such events we do not have a contradiction to the result that the normal distribution is indeed an excellent approximation to the true distribution of men, which is itself not precisely known.

Exercises for Chapter 16

16.1 Let $a \in \mathbf{R}^n$, $a \neq 0$, and $\mu \in \mathbf{R}^n$. Let H be the hyperplane in $\mathbf{R}^n$ given by

$$H = \{x \in \mathbf{R}^n : \langle x - \mu, a \rangle = 0\}.$$

Show that $m_n(H) = 0$ where m_n is n-dimensional Lebesgue measure, and deduce that

$$\int_H f(x)dx = \int_{-\infty}^{\infty} \ldots \int_{\infty}^{\infty} f(x_1, \ldots, x_n) 1_H(X_1, \ldots, x_n) dx_1 \ldots dx_n = 0$$

for any Borel function f on $\mathbf{R}^n$.

16.2 Let $\mathcal{L}(Y) = N(0, 1)$, and let $a > 0$. Let

$$Z = \begin{cases} Y & \text{if } |Y| \leq a, \\ -Y & \text{if } |Y| > a. \end{cases}$$

Show that $\mathcal{L}(Z) = N(0, 1)$ as well.

16.3 Let X be $N(0, 1)$ and let Z be independent of X with $P(Z = 1) = P(Z = -1) = \frac{1}{2}$. Let $Y = ZX$. Show $\mathcal{L}(Y) = N(0, 1)$, but that (X, Y) is not Gaussian (i.e., not Multivariate Normal).

16.4 Let (X, Y) be Gaussian with mean (μ_X, μ_Y) and covariance matrix Q and $\det(Q) > 0$. Let ρ be the correlation coefficient

$$\rho = \frac{\text{Cov}(X, Y)}{\sqrt{\text{Var}(X)\text{Var}(Y)}}.$$

Show that if $-1 < \rho < 1$ the density of (X, Y) exists and is equal to:

$$f_{(X,Y)}(x, y) = \frac{1}{2\pi\sigma_X\sigma_Y\sqrt{1-\rho^2}} \exp\left\{\frac{-1}{2(1-\rho^2)} \left(\left(\frac{x - \mu_X}{\sigma_X}\right)^2\right.\right.$$
$$\left.\left. -\frac{2\rho(x - \mu_X)(y - \mu_Y)}{\sigma_X\sigma_Y} + \left(\frac{y - \mu_Y}{\sigma_Y}\right)^2\right)\right\}.$$

Show that if $\rho = -1$ or $\rho = 1$, then the density of (X, Y) does not exist.

16.5 Let ρ be in between -1 and 1, and μ_j, σ_j^2 ($j = 1, 2$) be given. Construct X_1, X_2 Normals with means μ_1, μ_2; variances σ_1^2, σ_2^2; and correlation ρ. (*Hint:* Let Y_1, Y_2 be i.i.d. $N(0, 1)$ and set $U_1 = Y_1$ and $U_2 = \rho Y_1 + \sqrt{1 - \rho^2}\, Y_2$. Then let $X_j = \mu_j + \sigma_j Y_j$ ($j = 1, 2$).)

16.6 Suppose X is Gaussian $N(\mu, Q)$ on $\mathbf{R}^n$, with $\det(Q) > 0$. Show that there exists a matrix B such that $Y = B(X - \mu)$ has the $N(0, I)$ distribution, where I is the $n \times n$ identity matrix. *(Special Note: This shows that any Gaussian r.v. with non-degenerate covariance matrix can be linearly transformed into a standard normal.)*

16.7 Let X be Gaussian and let

$$Y = \sum_{j=1}^{n} a_j X_j,$$

where $X = (X_1, \ldots, X_n)$. Show that Y is univariate $N(\mu, \sigma^2)$ where

$$\mu = \sum_{j=1}^{n} a_j E\{X_j\}$$

and

$$\sigma^2 = \sum_{j=1}^{n} a_j^2 \mathrm{Var}(X_j) + 2 \sum_{j<k} a_j a_k \mathrm{Cov}\,(X_j, X_k).$$

16.8 Let (X, Y) be bivariate normal $N(\mu, Q)$, where

$$Q = \begin{pmatrix} \sigma_X^2 & \rho \sigma_X \sigma_Y \\ \rho \sigma_X \sigma_Y & \sigma_Y^2 \end{pmatrix}$$

and ρ is the correlation coefficient ($|\rho| < 1$), ($\det(Q) > 0$). Then (X, Y) has a density f and show that its conditional density $f_{X=x}(y)$ is the density of a univariate normal with mean $\mu_Y + \rho \frac{\sigma_Y}{\sigma_X} (x - \mu_X)$ and variance $\sigma_Y^2 (1 - \rho^2)$. (cf. Theorem 12.2.)

16.9 Let X be $N(\mu, Q)$ with $\mu = (1, 1)$ and $Q = \begin{pmatrix} 3 & 1 \\ 1 & 2 \end{pmatrix}$. Find the conditional distribution of $Y = X_1 + X_2$ given $Z = X_1 - X_2 = 0$.

$$\left[\text{Answer: } f_{Z=0}(y) = \frac{1}{\sqrt{2\pi}\sqrt{\frac{20}{3}}} \exp\left\{ -\frac{1}{2} \frac{(y-2)^2}{\frac{20}{3}} \right\}. \right]$$

16.10 Let $\mathcal{L}(X) = N(\mu, Q)$ with $\det(Q) > 0$. Show that the conditional distributions of any number of coordinates of X, knowing the others, are also multivariate normal (cf. Theorem 12.2). [This Exercise generalizes Exercise 16.8.]

16.11 (Gut, 1995). Let (X, Y) have joint density

$$f_{(X,Y)}(x, y) = c \exp\left\{ -(1 + x^2)(1 + y^2) \right\}, \qquad -\infty < x, y < \infty,$$

where c is chosen so that f is a density. Show that f is *not* the density of a bivariate normal but that $f_{X=x}(y)$ and $f_{Y=y}(x)$ are each normal densities. (This shows that the converse of Exercise 16.10 does not hold.)

16.12 Let (X, Y) be Bivariate Normal with correlation coefficient ρ and mean $(0, 0)$. Show that if $|\rho| < 1$, then $Z = \frac{X}{Y}$ is Cauchy with parameters $\alpha = \rho\frac{\sigma_X}{\sigma_Y}$ and $\beta = \frac{\sigma_X}{\sigma_Y}\sqrt{1 - \rho^2}$. (Note: This result was already established in Example 12.5 when X and Y were independent.) We conclude that *the ratio of two centered Bivariate Normals is a Cauchy r.v.*

16.13 * Let (X, Y) be bivariate normal with mean 0 and correlation coefficient ρ. Let β be such that

$$\cos\beta = \rho \quad (0 \le \beta \le \pi)$$

and show that

$$P\{XY < 0\} = \frac{\beta}{\pi}.$$

(*Hint:* Recall from Exercise 16.12 that if $Z = \frac{X}{Y}$ and $z = \rho\frac{\sigma_X}{\sigma_Y}$, then

$$F_Z(z) = \frac{1}{2} + \frac{1}{\pi}\text{Arctan}\left(\frac{z\sigma_Y - \rho\sigma_X}{\sigma_X\sqrt{1 - \rho^2}}\right).$$

Let $\alpha = \text{Arcsin}\,\rho$ $(-\frac{\pi}{2} \le \alpha \le \frac{\pi}{2})$ and show first $P(XY < 0) = \frac{1}{2} - \frac{\alpha}{\pi}$, using that $\text{Arctan}\,\frac{\rho}{\sqrt{1-\rho^2}} = \text{Arcsin}\,\rho$.)

16.14 Let (X, Y), α and ρ be as in Exercise 16.13. Show that

$$P\{X > 0, Y > 0\} = P\{X < 0, Y < 0\} = \frac{1}{4} + \frac{\alpha}{2\pi};$$

$$P\{X > 0, Y < 0\} = P\{X < 0, Y > 0\} = \frac{1}{4} - \frac{\alpha}{2\pi}.$$

16.15 * Let (X, Y) be bivariate normal with density

$$f_{(X,Y)}(x, y) = \frac{1}{2\pi\sigma_X\sigma_Y\sqrt{1 - \rho^2}}e^{-\frac{1}{2(1-\rho^2)}\left(\frac{x^2}{\sigma_X^2} - \frac{2\rho xy}{\sigma_X\sigma_Y} + \frac{y^2}{\sigma_Y^2}\right)}.$$

Show that:

a) $E\{XY\} = \rho\sigma_X\sigma_Y$
b) $E\{X^2Y^2\} = E\{X^2\}E\{Y^2\} + 2(E\{XY\})^2$
c) $E\{|XY|\} = \frac{2\sigma_X\sigma_Y}{\pi}(\cos\alpha + \alpha\sin\alpha)$ where α is given by $\sin\alpha = \rho$ $(-\frac{\pi}{2} \le \alpha \le \frac{\pi}{2})$ (cf. Exercise 16.13).

16.16 Let (X, Y) be bivariate normal with correlation ρ and $\sigma_X^2 = \sigma_Y^2$. Show that X and $Y - \rho X$ are independent.

16.17 Let X be $N(\mu, Q)$ with $\det(Q) > 0$, with X $\mathbf{R}^n$-valued. Show that

$$(X - \mu)^*Q^{-1}(X - \mu) \quad \text{is} \quad \chi^2(n).$$

16.18 Let $X_1, \ldots, X_n$ be i.i.d. $N(0, \sigma^2)$, and let

$$\overline{x} = \frac{1}{n} \sum_{j=1}^{n} X_j \text{ and } S^2 = \frac{1}{n-1} \sum_{j=1}^{n} (X_j - \overline{x})^2.$$

Recall from Exercise 15.13 that $\overline{x}$ and S^2 are independent. Show that

$$\sum_{j=1}^{n} X_j^2 = \sum_{j=1}^{n} (X_j - \overline{x})^2 + n\overline{x}^2$$

and deduce that $(n-1)S^2/\sigma^2$ has a χ_{n-1}^2 distribution and that $n\overline{x}^2/\sigma^2$ has a χ_1^2 distribution.

16.19 Let $\varepsilon_1, \ldots, \varepsilon_n$ be i.i.d. $N(0, \sigma^2)$ and suppose $Y_i = \alpha + \beta x_i + \varepsilon_i$, $1 \leq i \leq n$. Suppose also that all x_i's are not equal, and set $\overline{x} = \frac{1}{n} \sum_{i=1}^{n} x_i$. We define *regression residuals* to be

$$\hat{\varepsilon}_i = Y_i - A - B x_i,$$

where A and B are given in (16.16).

a) Show that $E\{\hat{\varepsilon}_i\} = 0$, $1 \leq i \leq n$
b)* Show that

$$\text{Var}(\hat{\varepsilon}_i) = \sigma^2 \left(\frac{n-1}{n} - \frac{(x_i - \overline{x})^2}{\sum_{i=1}^{n} (x_i - \overline{x})^2} \right)$$

16.20 Let $\varepsilon_1, \ldots, \varepsilon_n$ and $\hat{\varepsilon}_1, \ldots, \hat{\varepsilon}_n$ be as in Exercise 16.19. Suppose σ is unknown, and define

$$\hat{\sigma}^2 = \frac{1}{n} \sum_{i=1}^{n} \hat{\varepsilon}_i^2.$$

Show that $E\{\hat{\sigma}^2\} = \frac{n-2}{n}\sigma^2$. (Since $E\{\hat{\sigma}^2\} \neq \sigma^2$, $\hat{\sigma}^2$ is said to be a *biased* estimator for σ^2; an unbiased estimator for σ^2 is $S^2 = \frac{n}{n-2}\hat{\sigma}^2$.)

16.21 Let $\varepsilon_1, \ldots, \varepsilon_n$, A, B, S^2 be as in Exercises 16.19 and 16.20. Show that (A, B) and S^2 are independent.

17 Convergence of Random Variables

In elementary mathematics courses (such as Calculus) one speaks of the convergence of functions: $f_n : \mathbf{R} \to \mathbf{R}$, then $\lim_{n \to \infty} f_n = f$ if $\lim_{n \to \infty} f_n(x) = f(x)$ for all x in $\mathbf{R}$. This is called *pointwise convergence of functions*. A random variable is of course a function ($X : \Omega \to \mathbf{R}$ for an abstract space Ω), and thus we have the same notion: a sequence $X_n : \Omega \to \mathbf{R}$ *converges pointwise to X if* $\lim_{n \to \infty} X_n(\omega) = X(\omega)$, for all $\omega \in \Omega$. This natural definition is surprisingly useless in probability. The next example gives an indication why.

Example 1: Let X_n be an i.i.d. sequence of random variables with $P(X_n = 1) = p$ and $P(X_n = 0) = 1 - p$. For example we can imagine tossing a slightly unbalanced coin (so that $p > \frac{1}{2}$) repeatedly, and $\{X_n = 1\}$ corresponds to heads on the nth toss and $\{X_n = 0\}$ corresponds to tails on the nth toss. In the "long run", we would expect the proportion of heads to be p; this would justify our model that claims the probability of heads is p. Mathematically we would want

$$\lim_{n \to \infty} \frac{X_1(\omega) + \ldots + X_n(\omega)}{n} = p \quad \text{for all } \omega \in \Omega.$$

This simply does not happen! For example let $\omega_0 = \{T, T, T, \ldots\}$, the sequence of all tails. For this ω_0,

$$\lim_{n \to \infty} \frac{1}{n} \sum_{j=1}^{n} X_j(\omega_0) = 0.$$

More generally we have the event

$$A = \{\omega : \text{only a finite number of heads occur}\}.$$

Then

$$\lim_{n \to \infty} \frac{1}{n} \sum_{j=1}^{n} X_j(\omega) = 0 \text{ for all } \omega \in A.$$

We readily admit that the event A is very unlikely to occur. Indeed, we can show (Exercise 17.13) that $P(A) = 0$. In fact, what we will eventually show (see the Strong Law of Large Numbers [Chapter 20]) is that

$$P\left(\left\{\omega : \lim_{n\to\infty} \frac{1}{n} \sum_{j=1}^{n} X_j(\omega) = p\right\}\right) = 1.$$

This type of convergence of random variables, where we do not have convergence for *all* ω but do have convergence for *almost all* ω (i.e., the set of ω where we do have convergence has probability one), is what typically arises.

Caveat: In this chapter we will assume that all random variables are defined on a given, fixed probability space $(\Omega, \mathcal{A}, P)$ and takes values in $\mathbf{R}$ or $\mathbf{R}^n$. We also denote by $|x|$ the Euclidean norm of $x \in \mathbf{R}^n$.

Definition 17.1. *We say that a sequence of random variables $(X_n)_{n\geq 1}$ converges almost surely to a random variable X if*

$$N = \left\{\omega : \lim_{n\to\infty} X_n(\omega) \neq X(\omega)\right\} \text{ has } P(N) = 0.$$

Recall that the set N is called a null *set, or a* negligible *set.*

Note that

$$N^c = \Lambda = \left\{\omega : \lim_{n\to\infty} X_n(\omega) = X(\omega)\right\} \text{ and then } P(\Lambda) = 1.$$

We usually abbreviate almost sure convergence by writing

$$\lim_{n\to\infty} X_n = X \text{ a.s.}$$

We have given an example of almost sure convergence from coin tossing preceding this definition.

Just as we defined almost sure convergence because it naturally occurs when "pointwise convergence" (for all "points") fails, we need to introduce two more types of convergence. These next two types of convergence also arise naturally when a.s. convergence fails, and they are also useful as tools to help to show that a.s. convergence holds.

Definition 17.2. *A sequence of random variables $(X_n)_{n\geq 1}$ converges in L^p to X (where $1 \leq p < \infty$) if $|X_n|$, $|X|$ are in L^p and:*

$$\lim_{n\to\infty} E\{|X_n - X|^p\} = 0.$$

Alternatively one says X_n converges to X in p^{th} mean, and one writes

$$X_n \xrightarrow{L^p} X.$$

The most important cases for convergence in p^{th} mean are when $p = 1$ and when $p = 2$. When $p = 1$ and all r.v.'s are one-dimensional, we have

$|E\{X_n - X\}| \leq E\{|X_n - X|\}$ and $|E\{|X_n|\} - E\{|X|\}| \leq E\{|X_n - X|\}$ because $||x| - |y|| \leq |x - y|$. Hence

$$X_n \xrightarrow{L^1} X \quad \text{implies} \quad E\{X_n\} \to E\{X\} \quad \text{and} \quad E\{|X_n|\} \to E\{|X|\}. \quad (17.1)$$

Similarly, when $X_n \xrightarrow{L^p} X$ for $p \in (1, \infty)$, we have that $E\{|X_n|^p\}$ converges to $E\{|X|^p\}$: see Exercise 17.14 for the case $p = 2$.

Definition 17.3. *A sequence of random variables* $(X_n)_{n \geq 1}$ *converges in probability to* X *if for any* $\varepsilon > 0$ *we have*

$$\lim_{n \to \infty} P(\{\omega : |X_n(\omega) - X(\omega)| > \varepsilon\}) = 0.$$

This is also written

$$\lim_{n \to \infty} P(|X_n - X| > \varepsilon) = 0,$$

and denoted

$$X_n \xrightarrow{P} X.$$

Using the epsilon-delta definition of a limit, one could alternatively say that X_n tends to X in probability if for any $\varepsilon > 0$, any $\delta > 0$, there exists $N = N(\delta)$ such that

$$P(|X_n - X| > \varepsilon) < \delta$$

for all $n \geq N$.

Before we establish the relationships between the different types of convergence, we give a surprisingly useful small result which characterizes convergence in probability.

Theorem 17.1. $X_n \xrightarrow{P} X$ *if and only if*

$$\lim_{n \to \infty} E\left\{\frac{|X_n - X|}{1 + |X_n - X|}\right\} = 0.$$

Proof. There is no loss of generality by taking $X = 0$. Thus we want to show $X_n \xrightarrow{P} 0$ if and only if $\lim_{n \to \infty} E\{\frac{|X_n|}{1+|X_n|}\} = 0$. First suppose that $X_n \xrightarrow{P} 0$. Then for any $\varepsilon > 0$, $\lim_{n \to \infty} P(|X_n| > \varepsilon) = 0$. Note that

$$\frac{|X_n|}{1 + |X_n|} \leq \frac{|X_n|}{1 + |X_n|} 1_{\{|X_n| > \varepsilon\}} + \varepsilon 1_{\{|X_n| \leq \varepsilon\}} \leq 1_{\{|X_n| > \varepsilon\}} + \varepsilon.$$

Therefore

$$E\left\{\frac{|X_n|}{1 + |X_n|}\right\} \leq E\left\{1_{\{|X_n| > \varepsilon\}}\right\} + \varepsilon = P(|X_n| > \varepsilon) + \varepsilon.$$

Taking limits yields

$$\lim_{n\to\infty} E\left\{\frac{|X_n|}{1+|X_n|}\right\} \le \varepsilon;$$

since ε was arbitrary we have $\lim_{n\to\infty} E\{\frac{|X_n|}{1+|X_n|}\} = 0$.

Next suppose $\lim_{n\to\infty} E\{\frac{|X_n|}{1+|X_n|}\} = 0$. The function $f(x) = \frac{x}{1+x}$ is strictly increasing. Therefore

$$\frac{\varepsilon}{1+\varepsilon} 1_{\{|X_n|>\varepsilon\}} \le \frac{|X_n|}{1+|X_n|} 1_{\{|X_n|>\varepsilon\}} \le \frac{|X_n|}{1+|X_n|}.$$

Taking expectations and then limits yields

$$\frac{\varepsilon}{1+\varepsilon} \lim_{n\to\infty} P(|X_n| > \varepsilon) \le \lim_{n\to\infty} E\left\{\frac{|X_n|}{1+|X_n|}\right\} = 0.$$

Since $\varepsilon > 0$ is fixed, we conclude $\lim_{n\to\infty} P(|X_n| > \varepsilon) = 0$.   □

Remark: What this theorem says is that $X_n \xrightarrow{P} X$ iff[1] $E\{f(|X_n - X|)\} \to 0$ for the function $f(x) = \frac{|x|}{1+|x|}$. A careful examination of the proof shows that the same equivalence holds for any function f on $\mathbf{R}_+$ which is bounded, non-decreasing on $[0,\infty)$, continuous, and with $f(0) = 0$ and $f(x) > 0$ when $x > 0$. For example we have $X_n \xrightarrow{P} X$ iff $E\{|X_n - X| \wedge 1\} \to 0$ and also iff $E\{\arctan(|X_n - X|)\} \to 0$.

The next theorem shows that convergence in probability is the weakest of the three types of convergence (a.s., L^p, and probability).

Theorem 17.2. *Let $(X_n)_{n\ge1}$ be a sequence of random variables.*

a) *If $X_n \xrightarrow{L^p} X$, then $X_n \xrightarrow{P} X$.*
b) *If $X_n \xrightarrow{a.s.} X$, then $X_n \xrightarrow{P} X$.*

Proof. (a) Recall that for an event A, $P(A) = E\{1_A\}$, where 1_A is the indicator function of the event A. Therefore,

$$P\{|X_n - X| > \varepsilon\} = E\left\{1_{\{|X_n - X|>\varepsilon\}}\right\}.$$

Note that $\frac{|X_n - X|^p}{\varepsilon^p} > 1$ on the event $\{|X_n - X| > \varepsilon\}$, hence

$$\le E\left\{\frac{|X_n - X|^p}{\varepsilon^p} 1_{\{|X_n - X|>\varepsilon\}}\right\}$$

$$= \frac{1}{\varepsilon^p} E\left\{|X_n - X|^p 1_{\{|X_n - X|>\varepsilon\}}\right\},$$

and since $|X_n - X|^p \ge 0$ always, we can simply drop the indicator function to get:

---

[1] The notation *iff* is a standard notation shorthand for "if and only if"

$$\leq \frac{1}{\varepsilon^p} E\{|X_n - X|^p\}.$$

The last expression tends to 0 as n tends to ∞ (for fixed $\varepsilon > 0$), which gives the result.

(b) Since $\frac{|X_n - X|}{1 + |X_n - X|} \leq 1$ always, we have

$$\lim_{n \to \infty} E\left\{\frac{|X_n - X|}{1 + |X_n - X|}\right\} = E\left\{\lim_{n \to \infty} \frac{|X_n - X|}{1 + |X_n - X|}\right\} = E\{0\} = 0$$

by Lebesgue's Dominated Convergence Theorem (9.1(f)). We then apply Theorem 17.1. $\qquad \square$

The converse to Theorem 17.2 is not true; nevertheless we have two partial converses. The most delicate one concerns the relation with a.s. convergence, and goes as follows:

Theorem 17.3. *Suppose $X_n \xrightarrow{P} X$. Then there exists a subsequence n_k such that $\lim_{k \to \infty} X_{n_k} = X$ almost surely.*

Proof. Since $X_n \xrightarrow{P} X$ we have that $\lim_{n \to \infty} E\{\frac{|X_n - X|}{1 + |X_n - X|}\} = 0$ by Theorem 17.1. Choose a subsequence n_k such that $E\{\frac{|X_{n_k} - X|}{1 + |X_{n_k} - X|}\} < \frac{1}{2^k}$. Then $\sum_{k=1}^{\infty} E\{\frac{|X_{n_k} - X|}{1 + |X_{n_k} - X|}\} < \infty$ and by Theorem 9.2 we have that $\sum_{k=1}^{\infty} \frac{|X_{n_k} - X|}{1 + |X_{n_k} - X|} < \infty$ a.s.; since the general term of a convergent series must tend to zero, we conclude

$$\lim_{n_k \to \infty} |X_{n_k} - X| = 0 \text{ a.s.}$$

$\qquad \square$

Remark 17.1. Theorem 17.3 can also be proved fairly simply using the Borel–Cantelli Theorem (Theorem 10.5).

Example 2: $X_n \xrightarrow{P} X$ does not necessarily imply that X_n converges to X almost surely. For example take $\Omega = [0, 1]$, $\mathcal{A}$ the Borel sets on $[0, 1]$, and P the uniform probability measure on $[0, 1]$. (That is, P is just Lebesgue measure restricted to the interval $[0, 1]$.) Let A_n be any interval in $[0, 1]$ of length a_n, and take $X_n = 1_{A_n}$. Then $P(|X_n| > \varepsilon) = a_n$, and as soon as $a_n \to 0$ we deduce that $X_n \xrightarrow{P} 0$ (that is, X_n tends to 0 in probability). More precisely, let $X_{n,j}$ be the indicator of the interval $[\frac{j-1}{n}, \frac{j}{n}]$, $1 \leq j \leq n$, $n \geq 1$. We can make one sequence of the $X_{n,j}$ by ordering them first by increasing n, and then for each fixed n by increasing j. Call the new sequence Y_m. Thus the sequence would be:

$$X_{1,1}, X_{2,1}, X_{2,2}, X_{3,1}, X_{3,2}, X_{3,3}, X_{4,1}, \cdots$$
$$Y_1, \quad Y_2, \quad Y_3, \quad Y_4, \quad Y_5, \quad Y_6, \quad Y_7, \quad \cdots$$

Note that for each w and every n, there exists a j such that $X_{n,j}(w) = 1$. Therefore $\limsup_{m\to\infty} Y_m = 1$ a.s., while $\liminf_{m\to\infty} Y_m = 0$ a.s. Clearly then the sequence Y_m does not converge a.s. However Y_n is the indicator of an interval whose length a_n goes to 0 as $n \to \infty$, so the sequence Y_n does converge to 0 in probability.

The second partial converse of Theorem 17.2 is as follows:

Theorem 17.4. *Suppose $X_n \overset{P}{\to} X$ and also that $|X_n| \leq Y$, all n, and $Y \in L^p$. Then $|X|$ is in L^p and $X_n \overset{L^p}{\to} X$.*

Proof. Since $E\{|X_n|^p\} \leq E\{Y^p\} < \infty$, we have $X_n \in L^p$. For $\varepsilon > 0$ we have

$$\{|X| > Y + \varepsilon\} \subset \{|X| > |X_n| + \varepsilon\}$$
$$\subset \{|X| - |X_n| > \varepsilon\}$$
$$\subset \{|X - X_n| > \varepsilon\},$$

hence

$$P(|X| > Y + \varepsilon) \leq P(|X - X_n| > \varepsilon),$$

and since this is true for each n, we have

$$P(|X| > Y + \varepsilon) \leq \lim_{n\to\infty} P(|X - X_n| > \varepsilon) = 0,$$

by hypothesis. This is true for each $\varepsilon > 0$, hence

$$P(|X| > Y) \leq \lim_{m\to\infty} P(|X| > Y + \frac{1}{m}) = 0,$$

from which we get $|X| \leq Y$ a.s. Therefore $X \in L^p$ too.

Suppose now that X_n does not converge to X in L^p. There is a subsequence (n_k) such that $E\{|X_{n_k} - X|^p\} \geq \varepsilon$ for all k, and for some $\varepsilon > 0$. The subsequence X_{n_k} trivially converges to X in probability, so by Theorem 17.3 it admits a further subsequence $X_{n_{k_j}}$ which converges a.s. to X. Now, the r.v.'s $X_{n_{k_j}} - X$ tend a.s. to 0 as $j \to \infty$, while staying smaller than $2Y$, so by Lebesgue's Dominated Convergence we get that $E\{|X_{n_{k_j}} - X|^p\} \to 0$, which contradicts the property that $E\{|X_{n_k} - X|^p\} \geq \varepsilon$ for all k: hence we are done. □

The next theorem is elementary but also quite useful to keep in mind.

Theorem 17.5. *Let f be a continuous function.*

a) *If $\lim_{n\to\infty} X_n = X$ a.s., then $\lim_{n\to\infty} f(X_n) = f(X)$ a.s.*
b) *If $X_n \overset{P}{\to} X$, then $f(X_n) \overset{P}{\to} f(X)$.*

Proof. (a) Let $N = \{\omega : \lim_{n\to\infty} X_n(\omega) \neq X(\omega)\}$. Then $P(N) = 0$ by hypothesis. If $\omega \notin N$, then

$$\lim_{n\to\infty} f(X_n(\omega)) = f\left(\lim_{n\to\infty} X_n(\omega)\right) = f(X(\omega)),$$

where the first equality is by the continuity of f. Since this is true for any $\omega \notin N$, and $P(N) = 0$, we have the almost sure convergence.

(b) For each $k > 0$, let us set:

$$\{|f(X_n) - f(X)| > \varepsilon\} \subset \{|f(X_n) - f(X)| > \varepsilon, |X| \leq k\} \cup \{|X| > k\}. \quad (17.2)$$

Since f is continuous, it is uniformly continuous on any bounded interval. Therefore for our ε given, there exists a $\delta > 0$ such that $|f(x) - f(y)| \leq \varepsilon$ if $|x - y| \leq \delta$ for x and y in $[-k, k]$. This means that

$$\{|f(X_n) - f(X)| > \varepsilon, |X| \leq k\} \subset \{|X_n - X| > \delta, |X| \leq k\} \subset \{|X_n - X| > \delta\}.$$

Combining this with (17.2) gives

$$\{|f(X_n) - f(X)| > \varepsilon\} \subset \{|X_n - X| > \delta\} \cup \{|X| > k\}. \quad (17.3)$$

Using simple subadditivity $(P(A \cup B) \leq P(A) + P(B))$ we obtain from (17.3):

$$P\{|f(X_n) - f(X)| > \varepsilon\} \leq P(|X_n - X| > \delta) + P(|X| > k).$$

However $\{|X| > k\}$ tends to the empty set as k increases to ∞ so $\lim_{k\to\infty} P(|X| > k) = 0$. Therefore for $\gamma > 0$ we choose k so large that $P(|X| > k) < \gamma$. Once k is fixed, we obtain the δ of (17.3), and therefore

$$\lim_{n\to\infty} P\left(|f(X_n) - f(X)| > \varepsilon\right) \leq \lim_{n\to\infty} P\left(|X_n - X| > \delta\right) + \gamma = \gamma.$$

Since $\gamma > 0$ was arbitrary, we deduce the result.   □

Exercises for Chapter 17

17.1 Let $X_{n,j}$ be as given in Example 2. Let $Z_{n,j} = n^{\frac{1}{p}} X_{n,j}$. Let Y_m be the sequence obtained by ordering the $Z_{n,j}$ as was done in Example 2. Show that Y_m tends to 0 in probability but that $(Y_m)_{m \geq 1}$ does not tend to 0 in L^p, although each Y_n belongs to L^p.

17.2 Show that Theorem 17.5(b) is false in general if f is not assumed to be continuous. (*Hint:* Take $f(x) = 1_{\{0\}}(x)$ and the X_n's tending to 0 in probability.)

17.3 Let X_n be i.i.d. random variables with $P(X_n = 1) = \frac{1}{2}$ and $P(X_n = -1) = \frac{1}{2}$. Show that

$$\frac{1}{n} \sum_{j=1}^{n} X_j$$

converges to 0 in probability. (*Hint:* Let $S_n = \sum_{j=1}^{n} X_j$, and use Chebyshev's inequality on $P\{|S_n| > n\varepsilon\}$.)

17.4 Let X_n and S_n be as in Exercise 17.3. Show that $\frac{1}{n^2} S_{n^2}$ converges to zero a.s. (*Hint:* Show that $\sum_{n=1}^{\infty} P\{\frac{1}{n^2}|S_{n^2}| > \varepsilon\} < \infty$ and use the Borel-Cantelli Theorem.)

17.5 * Suppose $|X_n| \leq Y$ a.s., each n, $n = 1, 2, 3 \ldots$ Show that $\sup_n |X_n| \leq Y$ a.s. also.

17.6 Let $X_n \xrightarrow{P} X$. Show that the characteristic functions φ_{X_n} converge pointwise to φ_X (*Hint:* Use Theorem 17.4.)

17.7 Let $X_1, \ldots, X_n$ be i.i.d. Cauchy random variables with parameters $\alpha = 0$ and $\beta = 1$. (That is, their density is $f(x) = \frac{1}{\pi(1+x^2)}$, $-\infty < x < \infty$.) Show that $\frac{1}{n} \sum_{j=1}^{n} X_j$ also has a Cauchy distribution. (*Hint:* Use Characteristic functions: See Exercise 14.1.)

17.8 Let $X_1, \ldots, X_n, \ldots$ be i.i.d. Cauchy random variables with parameters $\alpha = 0$ and $\beta = 1$. Show that there is no constant γ such that $\frac{1}{n} \sum_{j=1}^{n} X_j \xrightarrow{P} \gamma$. (*Hint:* Use Exercise 17.7.) Deduce that there is no constant γ such that $\lim_{n \to \infty} \frac{1}{n} \sum_{j=1}^{n} X_j = \gamma$ a.s. as well.

17.9 Let $(X_n)_{n \geq 1}$ have finite variances and zero means (i.e., $\text{Var}(X_n) = \sigma_{X_n}^2 < \infty$ and $E\{X_n\} = 0$, all n). Suppose $\lim_{n \to \infty} \sigma_{X_n}^2 = 0$. Show X_n converges to 0 in L^2 and in probability.

17.10 Let X_j be i.i.d. with finite variances and zero means. Let $S_n = \sum_{j=1}^{n} X_j$. Show that $\frac{1}{n} S_n$ tends to 0 in both L^2 and in probability.

17.11 * Suppose $\lim_{n \to \infty} X_n = X$ a.s. and $|X| < \infty$ a.s. Let $Y = \sup_n |X_n|$. Show that $Y < \infty$ a.s.

17.12 * Suppose $\lim_{n \to \infty} X_n = X$ a.s. Let $Y = \sup_n |X_n - X|$. Show $Y < \infty$ a.s. (see Exercise 17.11), and define a new probability measure Q by

$$Q(A) = \frac{1}{c} E \left\{ 1_A \frac{1}{1+Y} \right\}, \quad \text{where } c = E \left\{ \frac{1}{1+Y} \right\}.$$

Show that X_n tends to X in L^1 under the probability measure Q.

17.13 Let A be the event described in Example 1. Show that $P(A) = 0$. (*Hint:* Let

$$A_n = \{ \text{ Heads on } n^{\text{th}} \text{ toss } \}.$$

Show that $\sum_{n=1}^{\infty} P(A_n) = \infty$ and use the Borel-Cantelli Theorem (Theorem 10.5.))

17.14 Let X_n and X be real-valued r.v.'s in L^2, and suppose that X_n tends to X in L^2. Show that $E\{X_n^2\}$ tends to $E\{X^2\}$ (*Hint:* use that $|x^2 - y^2| \leq (x - y)^2 + 2|y||x - y|$ and the Cauchy-Schwarz inequality).

17.15 * (Another *Dominated Convergence Theorem.*) Let $(X_n)_{n \geq 1}$ be random variables with $X_n \overset{P}{\to} X$ ($\lim_{n \to \infty} X_n = X$ in probability). Suppose $|X_n(\omega)| \leq C$ for a constant $C > 0$ and all ω. Show that $\lim_{n \to \infty} E\{|X_n - X|\} = 0$. (*Hint:* First show that $P(|X| \leq C) = 1$.)

18 Weak Convergence

In Chapter 17 we considered four types of convergence of random variables: pointwise everywhere, pointwise almost surely, convergence in p^{th} mean (L^p convergence), and convergence in probability. While all but the first differ from types of convergence seen in elementary Calculus courses, they are nevertheless squarely in the analysis tradition, and they can be thought of as variants of standard pointwise convergence. While these types of convergence are natural and useful in probability, there is yet another notion of convergence which is profoundly different from the four we have already seen. This convergence, known as *weak convergence*, is fundamental to the study of Probability and Statistics. As its name implies, it is a weak type of convergence. The weaker the requirements for convergence, the easier it is for a sequence of random variables to have a limit. What is unusual about weak convergence, however, is that the actual values of the random variables themselves are not important! It is simply the probabilities that they will assume those values that matter. That is, it is the *probability distributions of the random variables* that will be converging, and not the values of the random variables themselves. It is this difference that makes weak convergence a convergence of a different type than pointwise and its variants.

Since we will be dealing with the convergence of distributions of random variables, we begin by considering probability measures on $\mathbf{R}^d$, some $d \geq 1$.

Definition 18.1. *Let μ_n and μ be probability measures on $\mathbf{R}^d$ ($d \geq 1$). The sequence μ_n converges weakly to μ if $\int f(x)\mu_n(dx)$ converges to $\int f(x)\mu(dx)$ for each f which is real-valued, continuous and bounded on $\mathbf{R}^d$.*

At first glance this definition may look like it has a typographical error: one is used to considering

$$\lim_{n \to \infty} \int f_n(x)\mu(dx) = \int f(x)\mu(dx);$$

but here f remains fixed and it is indeed μ that varies. Note also that we do not consider all bounded Borel measurable functions f, but only the subset that are bounded and continuous.

Definition 18.2. *Let $(X_n)_{n \geq 1}$, X be $\mathbf{R}^d$-valued random variables. We say X_n converges in distribution to X (or equivalently X_n converges in law to X)*

if the distribution measures P^{X_n} converge weakly to P^X. We write $X_n \overset{D}{\to} X$, or equivalently $X_n \overset{\mathcal{L}}{\to} X$.

Theorem 18.1. *Let $(X_n)_{n \geq 1}$, X be $\mathbf{R}^d$-valued random variables. Then $X_n \overset{D}{\to} X$ if and only if*

$$\lim_{n \to \infty} E\{f(X_n)\} = E\{f(X)\},$$

for all continuous, bounded functions f on $\mathbf{R}^d$.

Proof. This is just a combination of Definitions 18.1 and 18.2, once we observe that

$$E\{f(X_n)\} = \int f(x) P^{X_n}(dx), \qquad E\{f(X)\} = \int f(x) P^X(dx).$$

$\square$

It is important to emphasize that if X_n converges in distribution to X, *there is no requirement that $(X_n)_{n \geq 1}$ and X be defined on the same probability space $(\Omega, \mathcal{A}, P)$!* Indeed in Statistics, for example, it happens that a sequence $(X_n)_{n \geq 1}$ all defined on one space will converge in distribution to a r.v. X that cannot exist on the same space the $(X_n)_{n \geq 1}$ were defined on! Thus the notion of weak convergence permits random variables to converge in ways that would otherwise be fundamentally impossible.

In order to have almost sure or L^p convergence, or convergence in probability, one always needs that $(X_n)_{n \geq 1}$, X are all defined on the same space. Thus *a priori* convergence in distribution is not comparable to the other kinds of convergence. Nevertheless, if by good fortune (or by construction) all of the $(X_n)_{n \geq 1}$ and X are all defined on the same probability space, then we can compare the types of convergence.

Theorem 18.2. *Let $(X_n)_{n \geq 1}$, X all be defined on a given and fixed probability space $(\Omega, \mathcal{A}, P)$. If X_n converges to X in probability, then X_n converges to X in distribution as well.*

Proof. Let f be bounded and continuous on $\mathbf{R}^d$. Then by Theorem 17.5 we know that $f(X_n)$ converges to $f(X)$ in probability too. Since f is bounded, $f(X_n)$ converges to $f(X)$ in L^1 by Theorem 17.4. Therefore $\lim_{n \to \infty} E\{f(X_n)\} = E\{f(X)\}$ by (17.1), and Theorem 18.1 gives the result. $\square$

There is a (very) partial converse to Theorem 18.2

Theorem 18.3. *Let $(X_n)_{n \geq 1}$, X be defined on a given fixed probability space $(\Omega, \mathcal{A}, P)$. If X_n converges to X in distribution, and if X is a r.v. equal a.s. to a constant, then X_n converges to X in probability as well.*

Proof. Suppose that X is a.s. equal to the constant a. The function $f(x) = \frac{|x-a|}{1+|x-a|}$ is bounded and continuous; therefore $\lim_{n\to\infty} E\{\frac{|X_n-a|}{1+|X_n-a|}\} = 0$, and hence X_n converges to a in probability by Theorem 17.1.   □

It is tempting to think that convergence in distribution implies the following: that if $X_n \overset{D}{\to} X$ then $P(X_n \in A)$ converges to $P(X \in A)$ for all Borel sets A. *This is almost never true.* We do have $P(X_n \in A)$ converges to $P(X \in A)$ for some sets A, but these sets are quite special. This is related to the convergence (in the real valued case) of distribution functions: indeed, if X_n are real valued and $X_n \overset{D}{\to} X$, then if $F_n(x) = P(X_n \leq x)$ were to converge to $F(x) = P(X \leq x)$, we would need to have $P(X_n \in (-\infty, x])$ converge to $P(X \in (-\infty, x])$ for all $x \in \mathbf{R}$, and even this is not always true!

Let us henceforth assume that $(X_n)_{n\geq1}$, X are real valued random variables and that $(F_n)_{n\geq1}$, F are their respective distribution functions. The next theorem is rather difficult and can be skipped. We note that it is much simpler if we assume that F, the distribution of the limiting random variable X, is itself continuous. This suffices for many applications, but we include a proof of Theorem 18.4 for completeness. For this theorem, recall that the distribution function F of a r.v. is nondecreasing and right-continuous, and so it has left limits everywhere, that is $\lim_{y\to x, y<x} F(y) = F(x-)$ exists for all x (see Exercise 18.4).

Theorem 18.4. *Let $(X_n)_{n\geq1}$, X be real valued random variables.*

a) *If $X_n \overset{D}{\to} X$ then $\lim_{n\to\infty} F_n(x) = F(x)$ for all x in the dense subset of* **R** *given by $D = \{x : F(x-) = F(x)\}$. ($F_n(x) = P(X_n \leq x)$; D is sometimes called the set of continuity points of F.)*
b) *Suppose $\lim_{n\to\infty} F_n(x) = F(x)$ for all x in a dense subset of* **R**. *Then $X_n \overset{D}{\to} X$.*

Proof of (a): Assume $X_n \overset{D}{\to} X$. Let $D = \{x : F(x-) = F(x)\}$. Then D is a dense subset of **R** since its complement (the set of discontinuities of F) is at most countably infinite (see Exercises 18.4 and 18.5), and the complement of a countable set is always dense in **R**.

Let us fix $x \in \mathbf{R}$. For each integer $p \geq 1$ let us introduce the following bounded, continuous functions:

$$f_p(y) = \begin{cases} 1 & \text{if } y \leq x \\ p(x-y) + 1 & \text{if } x < y < x + \dfrac{1}{p} \\ 0 & \text{if } x + \dfrac{1}{p} \leq y \end{cases}$$

$$g_p(y) = \begin{cases} 1 & \text{if } y \le x - \dfrac{1}{p} \\ p(x-y) & \text{if } x - \dfrac{1}{p} < y < x \\ 0 & \text{if } x \le y. \end{cases}$$

Then

$$\lim_{n\to\infty} E\{f_p(X_n)\} = E\{f_p(X)\}, \qquad \lim_{n\to\infty} E\{g_p(X_n)\} = E\{g_p(X)\}$$

for each $p \ge 1$. Note further that

$$E\{g_p(X_n)\} \le F_n(x) \le E\{f_p(X_n)\}$$

and hence

$$E\{g_p(X)\} \le \liminf_{n\to\infty} F_n(x) \le \limsup_{n\to\infty} F_n(x) \le E\{f_p(X)\}, \qquad \text{each } p \ge 1.$$
(18.1)

Now $\lim_{p\to\infty} f_p(y) = 1_{(-\infty,x]}(y)$ and $\lim_{p\to\infty} g_p(y) = 1_{(-\infty,x)}(y)$, hence by Lebesgue's dominated convergence theorem (Theorem 9.1(f)) we have that

$$\lim_{p\to\infty} E\{f_p(X)\} = E\left\{1_{(-\infty,x]}(X)\right\}$$
$$= P(X \le x)$$
$$= F(x),$$

and similarly $\lim_{p\to\infty} E\{g_p(X)\} = F(x-)$ (the left limit of F at point x). Combining these and (18.1) gives

$$F(x-) \le \liminf_{n} F_n(x) \le \limsup_{n} F_n(x) \le F(x). \qquad (18.2)$$

Therefore if $x \in D$, since we have $F(x-) = F(x)$, we readily deduce that $F_n(x) \to F(x)$.

Proof of (b): Now we suppose that $\lim_{n\to\infty} F_n(x) = F(x)$ for all $x \in \Delta$, where Δ is a dense subset of **R**. Let f be a bounded, continuous function on **R** and take $\varepsilon > 0$. Let r, s be in Δ such that

$$P(X \notin (r,s]) = 1 - F(s) + F(r) \le \varepsilon.$$

(Such r and s exist, since $F(x)$ decreases to 0 as x decreases to $-\infty$, and increases to 1 as x increases to $+\infty$, and since Δ is dense). Since F_n converges to F on Δ by hypothesis, there exists an N_1 such that for $n \ge N_1$,

$$P(X_n \notin (r,s]) = 1 - F_n(s) + F_n(r) \le 2\varepsilon. \qquad (18.3)$$

Since $[r,s]$ is a closed (compact) interval and f is continuous, we know f is uniformly continuous on $[r,s]$; hence there exists a *finite* number of points $r = r_0 < r_1 < \ldots < r_k = s$ such that

$$|f(x) - f(r_j)| \leq \varepsilon \quad \text{if } r_{j-1} \leq x \leq r_j,$$

and each of the r_j are in Δ, $1 \leq j \leq k$. (That we may choose r_j in Δ follows from the fact that Δ is dense.)

Next we set

$$g(x) = \sum_{j=1}^{k} f(r_j) 1_{(r_{j-1}, r_j]}(x) \tag{18.4}$$

and by the preceding we have $|f(x) - g(x)| \leq \varepsilon$ on $(r, s]$. Therefore if $\alpha = \sup_x |f(x)|$, we obtain

$$|E\{f(X_n)\} - E\{g(X_n)\}| \leq \alpha P(X_n \notin (r, s]) + \varepsilon, \tag{18.5}$$

and the same holds for X in place of X_n.

Using the definition (18.4) for g, observe that

$$E\{g(X_n)\} = \sum_{j=1}^{k} f(r_j)\{F_n(r_j) - F_n(r_{j-1})\}$$

and analogously

$$E\{g(X)\} = \sum_{j=1}^{k} f(r_j)\{F(r_j) - F(r_{j-1})\}.$$

Since all the r_j's are in Δ, we have $\lim_{n \to \infty} F_n(r_j) = F(r_j)$ for each j. Since there are only a finite number of r_j's, we know there exists an N_2 such that for $n \geq N_2$,

$$|E\{g(X_n)\} - E\{g(X)\}| \leq \varepsilon. \tag{18.6}$$

Let us now combine (18.5) for X_n and X and (18.6): if $n \geq \max(N_1, N_2)$, then

$$
\begin{aligned}
&|E\{f(X_n)\} - E\{f(X)\}| \\
&\leq |E\{f(X_n)\} - E\{g(X_n)\}| + |E\{g(X_n)\} - E\{g(X)\}| + |E\{g(X)\} \\
&\quad - E\{f(X)\}| \\
&\leq (\alpha P(X_n \notin (r, s]) + \varepsilon) + \varepsilon + (\alpha P(X \notin (r, s]) + \varepsilon) \\
&\leq (2\alpha\varepsilon + \varepsilon) + \varepsilon + (\alpha\varepsilon + \varepsilon) \\
&\leq 3\alpha\varepsilon + 3\varepsilon.
\end{aligned}
$$

Since ε was arbitrary, we conclude that $\lim_{n \to \infty} E\{f(X_n)\} = E\{f(X)\}$ for all bounded, continuous f; hence by Theorem 18.1 we have the result. $\square$

Examples:

1. Suppose that $(\mu_n)_{n\geq 1}$ is a sequence of probability measures on $\mathbf{R}$ that are all point masses (or, Dirac measures): that is, for each n there exists a point α_n such that $\mu_n(\{\alpha_n\}) = 1$ and

$$\mu_n(\{\alpha_n\}^c) = \mu_n(\mathbf{R} \setminus \{\alpha_n\}) = 0.$$

Then μ_n converges weakly to a limit μ if and only if α_n converges to a point α; and in this case μ is point mass at α. [*Special note::* "point mass" probability measures are usually written ε_α or δ_α in the literature, which is used to denote point mass of size one at the point α.] Note that $F_n(x) = 1_{[\alpha_n,\infty)}(x)$, and therefore $\lim_{n\to\infty} F_n(x) = F(x)$ on a dense subset easily implies that F must be of the form $1_{[\alpha,\infty)}(x)$, where $\alpha = \lim_{n\to\infty} \alpha_n$.

2. Let

$$F_n(x) = \begin{cases} 0 & \text{if } x \leq -\dfrac{1}{n} \\ \dfrac{1}{2} + \dfrac{n}{2}x & \text{if } -\dfrac{1}{n} < x < \dfrac{1}{n} \\ 1 & \text{if } x \geq \dfrac{1}{n}. \end{cases}$$

Then

$$\lim_{n\to\infty} F_n(x) = F(x) = 1_{[0,\infty)}(x)$$

for all x except $x = 0$; thus the set D of Theorem 18.4 is $D = \mathbf{R} \setminus \{0\}$. Thus if $\mathcal{L}(X_n)$ is given by F_n, then we have $X_n \overset{D}{\to} X$, where X is constant and equal to 0 a.s. ($\mathcal{L}(X)$ is given by F.) What we have shown is that a sequence of uniform random variables $(X_n)_{n\geq 1}$, with X_n uniform on $(-\frac{1}{n}, \frac{1}{n})$, converge weakly to 0 (i.e., the constant random variable equal to 0 a.s.).

3. Let $(X_n)_{n\geq 1}$, X be random variables with densities $f_n(x)$, $f(x)$. Then the distribution function

$$F(x) = \int_{-\infty}^{x} f(u)du$$

is continuous; thus $F(x-) = F(x)$ on all of $\mathbf{R}$. Suppose $f_n(x) \leq g(x)$, all n and x, and $\int_{-\infty}^{\infty} g(x)dx < \infty$, and $\lim_{n\to\infty} f_n(x) = f(x)$ almost everywhere. Then $F_n(x)$ converges to $F(x)$ by Lebesgue's dominated convergence theorem and thus $X_n \overset{D}{\to} X$.

Note that alternatively in this example we have that

$$\lim_{n\to\infty} \int h(x) P^{X_n}(dx) = \lim_{n\to\infty} \int h(x) f_n(x) dx$$

$$= \int h(x) \lim_n f_n(x) dx$$

$$= \int h(x) f(x) dx = \int h(x) P^X(dx)$$

for any bounded continuous function h by Lebesgue's dominated convergence theorem, and we have another proof that $X_n \overset{D}{\to} X$. This proof works also in the multi-dimensional case, and we see that a slightly stronger form of convergence than weak convergence takes place here: we need h above to be bounded and measurable, but the continuity is superfluous.

The previous example has the following extension, which might look a bit surprising: we can interchange limits and integrals in a case where the sequence of functions is not dominated by a single integrable function; this is due to the fact that all functions f_n and f below have integrals equal to 1.

Theorem 18.5. *Let $(X_n)_{n\geq 1}$, X be r.v.'s with values in $\mathbf{R}^d$, with densities f_n, f. If the sequence f_n converges pointwise (or even almost everywhere) to f, then $X_n \overset{D}{\to} X$.*

Proof. Let h be a bounded measurable function on $\mathbf{R}^d$, and $\alpha = \sup_x |h(x)|$. Put $h_1(x) = h(x) + \alpha$ and $h_2(x) = \alpha - h(x)$. These two functions are positive, and thus so are $h_1 f_n$ and $h_2 f_n$, all n. Since further for $i = 1, 2$ the sequence $h_i f_n$ converges almost everywhere to $h_i f$, we can apply Fatou's Lemma (see Theorem 9.1) to obtain

$$E\{h_i(X)\} = \int f(x)h_i(x)dx \leq \liminf_{n\to\infty} \int f_n(x)h_i(x)dx = \liminf_{n\to\infty} E\{h_i(X_n)\}. \tag{18.7}$$

Observe that $E\{h(X_n)\} = E\{h_1(X_n)\} - \alpha$ and $E\{h(X_n)\} = \alpha - E\{h_2(X_n)\}$, and the same equalities hold with X in place of X_n. Since $\liminf(x_n) = -\limsup(-x_n)$, it follows from (18.7) applied successively to $i = 1$ and $i = 2$ that

$$E\{h(X)\} \leq \liminf_{n\to\infty} E\{h(X_n)\},$$
$$E\{h(X)\} \geq \limsup_{n\to\infty} E\{h(X_n)\}.$$

Hence $E\{h(X_n)\}$ converges to $E\{h(X)\}$, and the theorem is proved. □

The next theorem is a version of what is known as "Helly's selection principle". It is a difficult theorem, but we will need it to establish the relation between weak convergence and convergence of characteristic functions. The condition (18.8), that the measures can be made arbitrarily small, uniformly in n, on the complement of a compact set, is often called *tightness*.

Theorem 18.6. *Let $(\mu_n)_{n\geq 1}$ be a sequence of probability measures on $\mathbf{R}$ and suppose*

$$\lim_{m\to\infty} \sup_n \mu_n([-m, m]^c) = 0. \tag{18.8}$$

Then there exists a subsequence n_k such that $(\mu_{n_k})_{k\geq 1}$ converge weakly.

Proof. Let $F_n(x) = \mu_n((-\infty, x])$. Note that for each $x \in \mathbf{R}$, $0 \leq F_n(x) \leq 1$ for all n, thus $(F_n(x))_{n \geq 1}$ is a bounded sequence of real numbers. Hence by the Bolzano-Weierstrass theorem there always exists a subsequence n_k such that $(F_{n_k}(x))_{k \geq 1}$ converges. (Of course the subsequence n_k *a priori* depends on the point x).

We need to construct a limit in a countable fashion, so we restrict our attention to the rational numbers in $\mathbf{R}$ (denoted $\mathbf{Q}$). Let $r_1, r_2, \ldots, r_j, \ldots$ be an enumeration of the rationals. For r_1, there exists a subsequence $n_{1,k}$ of n such that the limit exists. We set:

$$G(r_1) = \lim_{k \to \infty} F_{n_{1,k}}(r_1).$$

For r_2, there exists a sub-subsequence $n_{2,k}$ such that the limit exists. Again, set:

$$G(r_2) = \lim_{k \to \infty} F_{n_{2,k}}(r_2).$$

That is, $n_{2,k}$ is a subsequence of $n_{1,k}$. We continue this way: for r_j, let $n_{j,k}$ be a subsequence of $n_{j-1,k}$ such that the limit exists. Again, set:

$$G(r_j) = \lim_{k \to \infty} F_{n_{j,k}}(r_j).$$

We then form just one subsequence by taking $n_k := n_{k,k}$. Thus for r_j, we have

$$G(r_j) = \lim_{k \to \infty} F_{n_k}(r_j),$$

since n_k is a subsequence of $n_{j,k}$ once $k \geq j$.

Next we set:

$$F(x) = \inf_{\substack{y \in \mathbf{Q} \\ y > x}} G(y). \tag{18.9}$$

Since the function G defined on $\mathbf{Q}$ is non-decreasing, so also is the function F given in (18.9), and it is right continuous by construction.

Let $\varepsilon > 0$. By hypothesis there exists an m such that

$$\mu_n([-m, m]^c) \leq \varepsilon$$

for all n simultaneously. Therefore

$$F_n(x) \leq \varepsilon \text{ if } x < -m, \text{ and } F_n(x) \geq 1 - \varepsilon \text{ if } x > m;$$

therefore we have the same for G, and finally

$$\left. \begin{array}{ll} F(x) \leq \varepsilon & \text{if } x < -m \\ F(x) \geq 1 - \varepsilon & \text{if } x \geq m. \end{array} \right\} \tag{18.10}$$

Since $0 \leq F \leq 1$, F is right continuous and non-decreasing, property (18.10) gives that F is a true distribution function, corresponding to a probability measure μ on $\mathbf{R}$.

Finally, suppose x is such that $F(x-) = F(x)$. For $\varepsilon > 0$, there exist $y, z \in \mathbf{Q}$ with $y < x < z$ and

$$F(x) - \varepsilon \le G(y) \le F(x) \le G(z) \le F(x) + \varepsilon.$$

Therefore for large enough k,

$$F(x) - 2\varepsilon \le F_{n_k}(y) \le F_{n_k}(x) \le F_{n_k}(z) \le F(x) + 2\varepsilon. \tag{18.11}$$

The inequalities (18.11) give that

$$F(x) - 2\varepsilon \le F(y) \le \liminf_{k\to\infty} F_{n_k}(x)$$
$$\le \limsup_{k\to\infty} F_{n_k}(x) \le F(z) \le F(x) + 2\varepsilon$$

and by the pinching theorem the $\liminf$ and $\limsup$ above must be equal and equal to $\lim_{k\to\infty} F_{n_k}(x) = F(x)$. Thus μ_{n_k} converges weakly to μ by Theorem 18.4. □

Remark 18.1. This theorem also has a multi-dimensional version (the proof is similar, but more complicated): let μ_n be probabilities on $\mathbf{R}^d$. Then it suffices to replace the condition (18.8) by

$$\lim_{m\to\infty} \sup_n \mu_n(\{x \in \mathbf{R}^d : |x| > m\}) = 0. \tag{18.12}$$

A useful observation is that in order to show weak convergence, one does not have to check that $\int f \, d\mu_n$ converges to $\int f \, d\mu$ for *all* bounded, continuous f, but only for a well chosen subset of them. We state the next result in terms of the convergence of random variables.

Theorem 18.7. *Let $(X_n)_{n\ge1}$ be a sequence of random variables ($\mathbf{R}$ or $\mathbf{R}^d$-valued). Then $X_n \overset{D}{\to} X$ if and only if $\lim_{n\to\infty} E\{g(X_n)\} = E\{g(X)\}$ for all bounded Lipschitz continuous functions g.*

Proof. A function g is Lipschitz continuous if there exists a constant k such that $|g(x) - g(y)| \le k\|x - y\|$, all x, y. Note that necessity is trivial, so we show sufficiency. We need to show $\lim_{n\to\infty} E\{f(X_n)\} = E\{f(X)\}$ for all bounded, continuous functions f. Let f be bounded continuous, and let $\alpha = \sup_x |f(x)|$. Suppose there exist Lipschitz continuous functions g_i, with $-\alpha \le g_i \le g_{i+1} \le f$, and $\lim_{i\to\infty} g_i(x) = f(x)$. Then

$$\liminf_{n\to\infty} E\{f(X_n)\} \ge \liminf_{n\to\infty} E\{g_i(X_n)\} = E\{g_i(X)\},$$

for each fixed i. But the Monotone Convergence Theorem applied to $g_i(X)+\alpha$ and $f(X) + \alpha$ implies

$$\lim_{i \to \infty} E\{g_i(X)\} = E\{f(X)\}.$$

Therefore

$$\liminf_{n \to \infty} E\{f(X_n)\} \geq E\{f(X)\}. \tag{18.13}$$

Next, exactly the same argument applied to $-f$ gives

$$\limsup_{n \to \infty} E\{f(X_n)\} \leq E\{f(X)\}, \tag{18.14}$$

and combining (18.13) and (18.14) gives

$$\lim_{n \to \infty} E\{f(X_n)\} = E\{f(X)\}.$$

It remains then only to construct the functions g_i. We need to find a sequence of Lipschitz functions $\{j_1, j_2, \ldots\}$ such that $\sup_k j_k(x) = f(x)$ and $j_k(x) \geq -\alpha$; then we can take $g_i(x) = \max\{j_1(x), \ldots, j_i(x)\}$, and we will be done.

By replacing $f(x)$ by $\tilde{f}(x) = f(x) + \alpha$ if necessary, without loss of generality we can assume the bounded function $f(x)$ is positive for all x. For each Borel set A define a function representing distance from A by

$$d_A(x) = \inf\{\|x - y\|; y \in A\}.$$

Then for rationals $r \geq 0$ and integers m, define

$$j_{m,r}(x) = r \wedge \left(m \, d_{\{y : f(y) \leq r\}}(x) \right).$$

Note that $|d_A(x) - d_A(y)| \leq \|x - y\|$ for any set A, hence $|j_{m,r}(x) - j_{m,r}(y)| \leq m\|x - y\|$, and so $j_{m,r}$ is Lipschitz continuous. Moreover $j_{m,r}(x) \leq r$, and it is zero if $f(x) \leq r$, so in particular $0 \leq j_{m,r}(x) \leq f(x)$.

Choose and fix a point x and $\varepsilon > 0$. Choose a positive rational r such that $f(x) - \varepsilon < r < f(x)$. Since f is continuous, $f(y) > r$ for all y in a neighborhood of x. Therefore $d_{\{y : f(y) \leq r\}}(x) > 0$, hence $j_{m,r}(x) = r > f(x) - \varepsilon$, for m sufficiently large. Since the rationals and integers are countable, the collection $\{j_{m,r}; m \in \mathbf{N}, r \in \mathbf{Q}_+\}$ is countable. If $\{j_i\}_{i \geq 1}$ represents an enumeration, we have seen that $\sup_i j_i(x) \geq f(x)$. Since $j_i \leq f$, each i, we have $\sup_i j_i(x) = f(x)$ and we are done. $\square$

Corollary 18.1. *Let $(X_n)_{n \geq 1}$ be a sequence of random variables ($\mathbf{R}$ or $\mathbf{R}^d$ valued). Then $X_n \xrightarrow{D} X$ if and only if $\lim_{n \to \infty} E\{g(X_n)\} = E\{g(X)\}$ for all bounded uniformly continuous functions g.*

Proof. If g is Lipschitz then it is uniformly continuous, so Theorem 18.7 gives the result. $\square$

Remark 18.2. In Theorem 18.7 we reduced the test class of functions for **R** or **R**^{d} valued random variables to converge weakly: we reduced it from bounded continuous functions to bounded Lipschitz continuous functions. One may ask if it can be further reduced. It can in fact be further reduced to $\mathcal{C}^\infty$ functions with compact support. See Exercises 18.19–18.22 in this regard, where the solutions to the exercises show that X_n converges to X in distribution if and only if $E\{f(X_n)\}$ converges to $E\{f(X)\}$ for all bounded, $\mathcal{C}^\infty$ functions f.

A consequence of Theorem 18.7 is *Slutsky's Theorem*, which is useful in Statistics.

Theorem 18.8 (Slutsky's Theorem). *Let $(X_n)_{n\geq 1}$ and $(Y_n)_{n\geq 1}$ be two sequences of **R**^{d} valued random variables, with $X_n \overset{\mathcal{D}}{\to} X$ and $\|X_n - Y_n\| \to 0$ in probability. Then $Y_n \overset{\mathcal{D}}{\to} X$.*

Proof. By Theorem 18.7 it suffices to show $\lim_n E\{f(Y_n)\} = E\{f(X)\}$ for all Lipschitz continuous, bounded f. Let then f be Lipschitz continuous. We have $|f(x) - f(y)| \leq k\|x - y\|$ for some real k, and $|f(x)| \leq \alpha$ for some real α. Then we have

$$\lim_{n\to\infty} |E\{f(X_n) - f(Y_n)\}| \leq \lim_{n\to\infty} E\{|f(X_n) - f(Y_n)|\}$$
$$\leq k\varepsilon + \lim_n E\{|f(X_n) - f(Y_n)|1_{\{\|X_n - Y_n\|>\varepsilon\}}\}.$$

But $\lim_{n\to\infty} E\{|f(X_n) - f(Y_n)|1_{\{\|X_n - Y_n\|>\varepsilon\}}\} \leq \lim_{n\to\infty} 2\alpha P\{\|X_n - Y_n\| > \varepsilon\} = 0$, and since $\varepsilon > 0$ is arbitrary we deduce that $\lim_{n\to\infty} |E\{f(X_n) - f(Y_n)\}| = 0$. Therefore

$$\lim_{n\to\infty} E\{f(Y_n)\} = \lim_{n\to\infty} E\{f(X_n)\} = E\{f(X)\},$$

and the theorem is proved.   □

We end this section with a consideration of the weak convergence of random variables that take on at most a countable number of values (e.g., the binomial, the Poisson, the hypergeometric, etc.). Since the state space is countable, we can assume that *every* function is continuous: this amounts to endowing the state space with the discrete topology (*Caution:* if the state space, say E, is naturally contained in **R** for example, then this discrete topology is induced by the usual topology on **R** only when the minimum of $|x - y|$ for $x, y \in E \cap [-m, m]$ is bounded away from 0 for all $m > 0$, like when $E = \mathbf{N}$ or $E = \mathbf{Z}$, where **Z** denotes the integer). The next theorem gives a simple characterization of weak convergence in this case, and it is comparable to Theorem 18.5.

Theorem 18.9. *Let X_n, X be random variables with at most countably many values. Then $X_n \overset{\mathcal{D}}{\to} X$ if and only if*

$$\lim_{n \to \infty} P(X_n = j) = P(X = j)$$

for each j in the state space of $(X_n)_{n \geq 1}$, X.

Proof. First suppose $X_n \overset{D}{\to} X$. Then

$$\lim_{n \to \infty} E\{f(X_n)\} = E\{f(X)\}$$

for every bounded, continuous function f (Theorem 18.1). Since all functions are continuous, choose $f(x) = 1_{\{j\}}(x)$ and we obtain the result.

Next, suppose $\lim_{n \to \infty} P(X_n = j) = P(X = j)$ for all j in the state space E. Let f be a bounded function with $\alpha = \sup_j |f(j)|$. Take $\varepsilon > 0$. Since

$$\sum_{j \in E} P(X = j) = 1$$

is a convergent series, there must exist a *finite* subset Λ of E such that

$$\sum_{j \in \Lambda} P(X = j) \geq 1 - \varepsilon;$$

also for n large enough we have as well:

$$\sum_{j \in \Lambda} P(X_n = j) \geq 1 - 2\varepsilon.$$

Note that

$$E\{f(X)\} = \sum_{j \in E} f(j)P(X = j),$$

so we have, for n large enough:

$$\left.\begin{aligned}\left|E\{f(X)\} - \sum_{j \in \Lambda} f(j)P(X = j)\right| &\leq \alpha\varepsilon \\ \left|E\{f(X_n)\} - \sum_{j \in \Lambda} f(j)P(X_n = j)\right| &\leq 2\alpha\varepsilon.\end{aligned}\right\} \tag{18.15}$$

Finally we note that since Λ is finite we have

$$\lim_{n \to \infty} \sum_{j \in \Lambda} f(j)P(X_n = j) = \sum_{j \in \Lambda} f(j)P(X = j). \tag{18.16}$$

Thus from (18.15) and (18.16) we deduce

$$\limsup_{n \to \infty} |E\{f(X_n)\} - E\{f(X)\}| \leq 3\alpha\varepsilon.$$

Since ε was arbitrary, we have

$$\lim_{n \to \infty} E\{f(X_n)\} = E\{f(X)\}$$

for each bounded (and *a fortiori* continuous) function f. Thus we have $X_n \overset{D}{\to} X$ by Theorem 18.1. □

Examples:

4. If μ_λ denotes the Poisson distribution with parameter λ, then

$$\mu_\lambda(j) = e^{-\lambda}\frac{\lambda^j}{j!},$$

   and thus if $\lambda_n \to \lambda$, we have $\mu_{\lambda_n}(j) \to \mu_\lambda(j)$ for each $j = 1, 2, 3, \ldots$ and by Theorem 18.9 we have that μ_{λ_n} converges weakly to μ_λ.

5. If μ_p denotes the Binomial $B(p, n)$ distribution and if $p_k \to p$, as in Example 4 and by Theorem 18.9 we have that μ_{p_k} converges weakly to μ_p.

6. Let $\mu_{n,p}$ denote the Binomial $B(p, n)$. Consider the sequence μ_{n,p_n} where $\lim_{n\to\infty} np_n = \lambda > 0$. Then as in Exercise 4.1 we have

$$\mu_{n,p_n}(k) = \frac{\lambda^k}{k!}\left(1 - \frac{\lambda}{n}\right)^n \left\{\frac{n}{n}\left(\frac{n-1}{n}\right)\cdots\left(\frac{n-k+1}{n}\right)\right\}\left(1 - \frac{\lambda}{n}\right)^{-k}$$

   for $0 \le k \le n$. Therefore for k fixed we have

$$\lim_{n\to\infty} \mu_{n,p_n}(k) = \frac{\lambda^k}{k!}e^{-\lambda},$$

   and hence by Theorem 18.9 we conclude that μ_{n,p_n} converges weakly to the Poisson distribution with parameter λ, where $\lambda = \lim_{n\to\infty} np_n$.

Exercises for Chapter 18

18.1 Show that if $X_n \xrightarrow{L^p} X$ $(p \geq 1)$, then $X_n \xrightarrow{\mathcal{D}} X$.

18.2 Let $\alpha \in \mathbf{R}^d$. Show by constructing it that there exists a continuous function $f : \mathbf{R}^d \to \mathbf{R}$ such that $0 \leq f(x) \leq 1$ for all $x \in \mathbf{R}^d$; $f(\alpha) = 0$; and $f(x) = 1$ if $|x - \alpha| \geq \varepsilon$ for a given $\varepsilon > 0$. (*Hint:* First solve this exercise when $d = 1$ and then mimic your construction for $d \geq 2$.)

18.3 Let X be a real valued random variable with distribution function F. Show that $F(x-) = F(x)$ if and only if $P(X = x) = 0$.

18.4 * Let $g : \mathbf{R} \to \mathbf{R}$, $0 \leq g(\alpha) \leq 1$, g nondecreasing, and suppose g is right continuous (that is, $\lim_{y \to x, y > x} g(y) = g(x)$ for all x). Show that g has left limits everywhere (that is, $\lim_{y \to x, y < x} g(y) = g(x-)$ exists for all x) and that the set $\Lambda = \{x : g(x-) \neq g(x)\}$ is at most countably infinite. (*Hint:* First show there are only a finite number of points x such that $g(x) - g(x-) > \frac{1}{k}$; then let k tend to ∞).

18.5 * Let F be the distribution function of a real valued random variable. Let $D = \{x : F(x-) = F(x)\}$ (notation of Exercise 18.4). Show that D is dense in $\mathbf{R}$. (*Hint:* Use Exercise 18.4 to show that the complement of D is at most countably infinite.)

18.6 Let $(X_n)_{n \geq 1}$ be a sequence of real valued random variables with $\mathcal{L}(X_n)$ uniform on $[-n, n]$. In what sense(s) do X_n converge to a random variable X? [*Answer:* None.]

18.7 Let $f_n(x)$ be densities on $\mathbf{R}$ and suppose $\lim_{n \to \infty} f_n(x) = e^{-x} 1_{(x>0)}$. If f_n is the density for a random variable X_n, each n, what can be said about the convergence of X_n as n tends to ∞? [*Answer:* $X_n \xrightarrow{\mathcal{D}} X$, where X is exponential with parameter 1.]

18.8 Let $(X_n)_{n \geq 1}$ be i.i.d. Cauchy with $\alpha = 0$ and $\beta = 1$. Let $Y_n = \frac{X_1 + \ldots + X_n}{n}$. Show that Y_n converges in distribution and find the limit. Does Y_n converge in probability as well?

18.9 Let $(X_n)_{n \geq 1}$ be a sequence of random variables and suppose $\sup_n E\{X_n^2\} < \infty$. Let μ_n be the distribution measure of X_n. Show that the sequence μ_n is tight (*Hint:* use Chebyshev's inequality).

18.10 * Let X_n, X and Y be real–valued r.v.'s, all defined on the same space $(\Omega, \mathcal{A}, P)$. Assume that

$$\lim_{n \to \infty} E\{f(X_n)g(Y)\} = E\{f(X)g(Y)\}$$

whenever f and g are bounded, and f is continuous, and g is Borel. Show that the sequence (X_n, Y) converges in law to (X, Y). If furthermore $X = h(Y)$ for some Borel function h, show that $X_n \xrightarrow{P} X$.

18.11 Let μ_α denote the Pareto (or Zeta) distribution with parameter α. Let $\alpha_n \to \alpha > 0$ and show that μ_{α_n} tends weakly to μ_α.

18.12 Let μ_α denote the Geometric distribution of parameter α. Let $\alpha_n \to \alpha > 0$, and show that μ_{α_n} tends weakly to μ_α.

18.13 Let $\mu_{(N,b,n)}$ be a Hypergeometric distribution, and let N go to ∞ in such a way that $p = \frac{b}{N}$ remains constant. The parameter n is held fixed. Show as N tends to ∞ as described above that $\mu_{(N,b,n)}$ converges weakly to the Binomial distribution $B(p, n)$.

18.14 (Slutsky's Theorem.) Let X_n converge in distribution to X and let Y_n converge in probability to a constant c. Show that (a) $X_n Y_n \xrightarrow{D} cX$ (in distribution) and (b) $\frac{X_n}{Y_n} \xrightarrow{D} \frac{X}{c}$ (in distribution), $(c \neq 0)$.

18.15 Let $(X_n)_{n\geq 1}$, $(Y_n)_{n\geq 1}$ all be defined on the same probability space. Suppose $X_n \xrightarrow{D} X$ and Y_n converges in probability to 0. Show that $X_n + Y_n$ converges in distribution to X.

18.16 Suppose real valued $(X_n)_{n\geq 1}$ have distribution functions F_n, and that $X_n \xrightarrow{D} X$. Let $p > 0$ and show that for every positive N,

$$\int_{-N}^{N} |x|^p F(dx) \leq \limsup_{n\to\infty} \int_{-N}^{N} |x|^p F_n(dx) < \infty.$$

18.17 * Let real valued $(X_n)_{n\geq 1}$ have distribution functions F_n, and X have distribution function F. Suppose for some $r > 0$,

$$\lim_{n\to\infty} \int_{-\infty}^{\infty} |F_n(x) - F(x)|^r dx = 0.$$

Show that $X_n \xrightarrow{D} X$. (*Hint:* Suppose there exists a continuity point y of F such that $\lim_{n\to\infty} F_n(y) \neq F(y)$. Then there exists $\varepsilon > 0$ and a subsequence $(n_k)_{k\geq 1}$ such that $|F_{n_k}(y) - F(y)| > \varepsilon$, all k. Show then $|F_{n_k}(x) - F(x)| > \frac{\varepsilon}{2}$ for either $x \in [y_1, y)$ or $x \in (y, y_2]$ for appropriate y_1, y_2. Use this to derive a contradiction.)

18.18 * Suppose a sequence $(F_n)_{n\geq 1}$ of distribution functions on $\mathbf{R}$ converges to a *continuous* distribution function F on $\mathbf{R}$. Show that the convergence is uniform in x $(-\infty < x < \infty)$. (*Hint:* Begin by showing there exist points $x_1, \ldots, x_m$ such that $F(x_1) < \varepsilon$, $F(x_{j+1}) - F(x_j) < \varepsilon$, and $1 - F(x_m) < \varepsilon$. Next show there exists N such that for $n > N$, $|F_n(x_j) - F(x_j)| < \varepsilon$, $1 \leq j \leq m$.)

18.19 Let f be uniformly continuous and X, Y two $\mathbf{R}$-valued random variables. Suppose that $|f(x) - f(y)| < \varepsilon$ whenever $|x - y| < \delta$. Show that

$$|E\{f(X)\} - E\{f(X + Y)\}| \leq \varepsilon + 2 \sup_x |f(x)| P\{|Y| \geq \delta\}.$$

18.20 * (Pollard [17]) Let $(X_n)_{n\geq 1}$, X, Y by **R**-valued random variables, all on the same space, and suppose that $X_n + \sigma Y$ converges in distribution to $X + \sigma Y$ for each fixed $\sigma > 0$. Show that X_n converges to X in distribution. (*Hint:* Use Exercise 18.19.)

18.21 (Pollard [17]) Let X and Y be independent r.v.'s on the same space, with values in **R** and assume Y is $N(0,1)$. Let f be bounded continuous. Show that

$$E\{f(X + \sigma Y)\} = E\{f_\sigma(X)\}$$

where

$$f_\sigma(x) = \frac{1}{\sqrt{2\pi}\sigma} \int_{-\infty}^{\infty} f(z) e^{-\frac{1}{2}|z-x|^2/\sigma^2} dz.$$

Show that f_σ is bounded and $\mathcal{C}^\infty$.

18.22 Let $(X_n)_{n\geq 1}$, X be **R**-valued random variables. Show that X_n converges to X in distribution if and only if $E\{f(X_n)\}$ converges to $E\{f(X)\}$ for all bounded $\mathcal{C}^\infty$ functions f. (*Hint:* Use Exercises 18.20 and 18.21.)

19 Weak Convergence and Characteristic Functions

Weak convergence is at the heart of much of probability and statistics. Limit theorems provide much of the justification of statistics, and they also have a myriad of other applications. There is an intimate relationship between weak convergence and characteristic functions, and it is indeed this relationship (provided by the next theorem) that makes characteristic functions so useful in the study of probability and statistics.

Theorem 19.1 (Lévy's Continuity Theorem). *Let $(\mu_n)_{n \geq 1}$ be a sequence of probability measures on $\mathbf{R}^d$, and let $(\hat{\mu}_n)_{n \geq 1}$ denote their Fourier transforms, or characteristic functions.*

a) *If μ_n converges weakly to a probability measure μ, then $\hat{\mu}_n(u) \to \hat{\mu}(u)$ for all $u \in \mathbf{R}^d$;*
b) *If $\hat{\mu}_n(u)$ converges to a function $f(u)$ for all $u \in \mathbf{R}^d$, and if in addition f is continuous at 0, then there exists a probability μ on $\mathbf{R}^d$ such that $f(u) = \hat{\mu}(u)$, and μ_n converges weakly to μ.*

Proof. (a) Suppose μ_n converges weakly to μ. Since e^{iux} is continuous and bounded in modulus,

$$\hat{\mu}_n(u) = \int e^{iux} \mu_n(dx)$$

converges to

$$\hat{\mu}(u) = \int e^{iux} \mu(dx)$$

by weak convergence (the function $x \mapsto e^{iux}$ is complex-valued, but we can consider separately the real-valued part $\cos(ux)$ and the imaginary part $\sin(ux)$, which are both bounded and continuous).

(b) Although we state the theorem for $\mathbf{R}^d$, we will give the proof only for $d = 1$. Suppose that $\lim_{n \to \infty} \hat{\mu}_n(u) = f(u)$ exists for all u. We begin by showing *tightness* (cf Theorem 18.6) of the sequence of probability measures μ_n. Using Fubini's theorem (Theorem 10.3 or more precisely Exercise 10.14) we have:

$$\int_{-\alpha}^{\alpha} \hat{\mu}_n(u) du = \int_{-\alpha}^{\alpha} \left\{ \int_{-\infty}^{\infty} e^{iux} \mu_n(dx) \right\} du$$

$$= \int_{-\infty}^{\infty} \left\{ \int_{-\alpha}^{\alpha} e^{iux} du \right\} \mu_n(dx);$$

and using that $e^{iux} = \cos(ux) + i\sin(ux)$,

$$= \int_{-\infty}^{\infty} \left\{ \int_{-\alpha}^{\alpha} \cos(ux) + i\sin(ux) du \right\} \mu_n(dx).$$

Since $\sin(ux)$ is an odd function, the imaginary integral is zero over the symmetric interval $(-\alpha, \alpha)$, and thus:

$$= \int_{-\infty}^{\infty} \frac{2}{x} \sin(\alpha x) \mu_n(dx).$$

Since $\int_{-\alpha}^{\alpha} 1 du = 2\alpha$, we have

$$\frac{1}{\alpha} \int_{-\alpha}^{\alpha} (1 - \hat{\mu}_n(u)) du = 2 - \int_{-\infty}^{\infty} \frac{2}{\alpha x} \sin(\alpha x) \mu_n(dx)$$

$$= 2 \int_{-\infty}^{\infty} \left(1 - \frac{\sin(\alpha x)}{\alpha x} \right) \mu_n(dx).$$

Now since $2(1 - \frac{\sin v}{v}) \geq 1$ if $|v| \geq 2$ and $2(1 - \frac{\sin v}{v}) \geq 0$ always, the above is

$$\geq \int_{-\infty}^{\infty} 1_{[-2,2]^c}(\alpha x) \mu_n(dx)$$

$$= \int 1_{[-2/\alpha, 2/\alpha]^c}(x) \mu_n(dx)$$

$$= \mu_n \left(\left[\frac{-2}{\alpha}, \frac{2}{\alpha} \right]^c \right).$$

Let $\beta = \frac{2}{\alpha}$ and we have the useful estimate:

$$\mu_n\left([-\beta, \beta]^c\right) \leq \frac{\beta}{2} \int_{-2/\beta}^{2/\beta} (1 - \hat{\mu}_n(u)) du. \qquad (19.1)$$

Let $\varepsilon > 0$. Since *by hypothesis f is continuous at 0*, there exists $\alpha > 0$ such that $|1 - f(u)| \leq \varepsilon/4$ if $|u| \leq 2/\alpha$. (This is because $\hat{\mu}_n(0) = 1$ for all n, whence $\lim_{n \to \infty} \hat{\mu}_n(0) = f(0) = 1$ as well.) Therefore,

$$\left| \frac{\alpha}{2} \int_{-2/\alpha}^{2/\alpha} (1 - f(u)) du \right| \leq \frac{\alpha}{2} \int_{-2/\alpha}^{2/\alpha} \frac{\varepsilon}{4} du = \frac{\varepsilon}{2}. \qquad (19.2)$$

Since $\hat{\mu}_n(u)$ are characteristic functions, $|\hat{\mu}_n(u)| \leq 1$, so by Lebesgue's dominated convergence theorem (Theorem 9.1 (f)) we have

$$\lim_{n \to \infty} \int_{-2/\alpha}^{2/\alpha} (1 - \hat{\mu}_n(u)) du = \int_{-2/\alpha}^{2/\alpha} (1 - f(u)) du.$$

Therefore there exists an N such that $n \geq N$ implies

$$\left| \int_{-2/\alpha}^{2/\alpha} (1 - \hat{\mu}_n(u)) du - \int_{-2/\alpha}^{2/\alpha} (1 - f(u)) du \right| \leq \frac{\varepsilon}{\alpha},$$

whence $\frac{\alpha}{2} \int_{-2/\alpha}^{2/\alpha} (1 - \hat{\mu}_n(u)) du \leq \varepsilon$. We next apply (19.1) to conclude $\mu_n([-\alpha, \alpha]^c) \leq \varepsilon$, for all $n \geq N$.

There are only a finite number of n before N, and for each $n < N$, there exists an α_n such that $\mu_n([-\alpha_n, \alpha_n]^c) \leq \varepsilon$. Let $a = \max(\alpha_1, \ldots, \alpha_n; \alpha)$. Then we have

$$\mu_n([-a, a]^c) \leq \varepsilon, \qquad \text{for all } n. \tag{19.3}$$

The inequality (19.3) above means that for the sequence $(\mu_n)_{n \geq 1}$, for any $\varepsilon > 0$ there exists an $a \in \mathbf{R}$ such that $\sup_n \mu_n([-a, a]^c) \leq \varepsilon$. Therefore we have shown:

$$\limsup_{m \to \infty} \sup_n \mu_n([-m, m]^c) = 0$$

for any fixed $m \in \mathbf{R}$.

We have established tightness for the sequence $\{\mu_n\}_{n \geq 1}$. We can next apply Theorem 18.6 to obtain a subsequence $(n_k)_{k \geq 1}$ such that μ_{n_k} converges weakly to μ as k tends to ∞. By part (a) of this theorem,

$$\lim_{k \to \infty} \hat{\mu}_{n_k}(u) = \hat{\mu}(u)$$

for all u, hence $f(u) = \hat{\mu}(u)$, and f is the Fourier transform of a probability measure.

It remains to show that the sequence $(\mu_n)_{n \geq 1}$ itself (and not just $(\mu_{n_k})_{k \geq 1}$) converges weakly to μ. We show this by the method of contradiction. Let F_n, F be distribution functions of μ_n and μ. That is,

$$F_n(x) = \mu_n((-\infty, x]); \quad F(x) = \mu((-\infty, x]).$$

Let D be the set of continuity points of F: that is,

$$D = \{x : F(x-) = F(x)\}.$$

Suppose that μ_n does not converge weakly to μ, then by Theorem 18.4 there must exist at least one point $x \in D$ and a subsequence $(n_k)_{k \geq 1}$ such that $\lim_{k \to \infty} F_{n_k}(x)$ exists (by taking a further subsequence if necessary) and moreover $\lim_{k \to \infty} F_{n_k}(x) = \beta \neq F(x)$. Next by Theorem 18.6 there also exists a subsequence of the subsequence (n_k) (that is, a sub-subsequence $(n_{k_j})_{j \geq 1}$), such that $(\mu_{n_{k_j}})_{j \geq 1}$ converges weakly to a limit ν as j tends to ∞. Exactly as we have argued, however, we get

$$\lim_{j \to \infty} \hat{\mu}_{n_{k_j}}(u) = \hat{\nu}(u),$$

and since $\lim \hat{\mu}_n(u) = f(u)$, we conclude $\hat{\nu}(u) = f(u)$. But we have seen that $f(u) = \hat{\mu}(u)$. Therefore by Theorem 14.1 we must have $\mu = \nu$. Finally, $\mu_{n_{k_j}}$ converging to $\nu = \mu$ implies (by Theorem 18.4) that $\lim_{j\to\infty} F_{n_{k_j}}(x) = F(x)$, since x is in D, the continuity set of μ, by hypothesis. But $\lim_{j\to\infty} F_{n_{k_j}}(x) = \beta \neq F(x)$, and we have a contradiction.   □

Remark 19.1. Actually more is true in Theorem 19.1a than we proved: one can show that if μ_n converges weakly to a probability measure μ on $\mathbf{R}^d$, then $\hat{\mu}_n$ converges to $\hat{\mu}$ *uniformly on compact subsets of* $\mathbf{R}^d$.

Example. Let $(X_n)_{n\geq 1}$ be a sequence of Poisson random variables with parameter $\lambda_n = n$. Then if

$$Z_n = \frac{1}{\sqrt{n}}(X_n - n), \qquad Z_n \xrightarrow{D} Z, \quad \text{where} \quad \mathcal{L}(Z) = N(0,1).$$

To see this, we have

$$E\left\{e^{iuZ_n}\right\} = E\left\{e^{iu\left(\frac{1}{\sqrt{n}}(X_n - n)\right)}\right\}$$

$$= e^{-iu\sqrt{n}}E\left\{e^{i\frac{u}{\sqrt{n}}X_n}\right\}$$

$$= e^{-iu\sqrt{n}}e^{n\left(e^{iu/\sqrt{n}}-1\right)}$$

by Example 13.3.

Continuing and using a Taylor expansion for e^z, we have the above equals

$$= e^{-iu\sqrt{n}}e^{n\left(i\frac{u}{\sqrt{n}} - \frac{u^2}{2n} - \frac{iu^3}{6n^{3/2}} + \cdots\right)}$$

$$= e^{-iu\sqrt{n}+iu\sqrt{n}}e^{-u^2/2}e^{-\frac{h(u,n)}{\sqrt{n}}}$$

$$= e^{-u^2/2}e^{-\frac{h(u,n)}{\sqrt{n}}}$$

where $h(u,n)$ stays bounded in n for each u and hence $\lim_{n\to\infty}\frac{h(u,n)}{\sqrt{n}} = 0$. Therefore,

$$\lim_{n\to\infty} \varphi_{Z_n}(u) = e^{-u^2/2},$$

and since $e^{-u^2/2}$ is the characteristic function of a $N(0,1)$, (Example 13.5), we have that Z_n converges weakly to Z by Theorem 19.1 b.

Exercises for Chapter 19

19.1 Let $(X_n)_{n\geq 1}$ be $N(\mu_n, \sigma_n^2)$ random variables. Suppose $\mu_n \to \mu \in \mathbf{R}$ and $\sigma_n^2 \to \sigma^2 \geq 0$. Show that $X_n \xrightarrow{D} X$, where $\mathcal{L}(X) = N(\mu, \sigma^2)$.

19.2 Let $(X_n)_{n\geq 1}$ be $N(\mu_n, \sigma_n^2)$ random variables. Suppose that $X_n \xrightarrow{D} X$ for some random variable X. Show that the sequences μ_n and σ_n^2 have limits $\mu \in \mathbf{R}$ and $\sigma^2 \geq 0$, and that X is $N(\mu, \sigma^2)$ (*Hint:* φ_{X_n} and φ_X being the characteristic functions of X_n and X, write $\varphi_{X_n} = e^{iu\mu_n - \frac{u^2 \sigma_n^2}{2}}$, and use Lévy's Theorem to obtain that $\varphi_X(u) = e^{iu\mu - \frac{u^2 \sigma^2}{2}}$ for some $\mu \in \mathbf{R}$ and $\sigma^2 \geq 0$).

19.3 Let $(X_n)_{n\geq 1}$, $(Y_n)_{n\geq 1}$ be sequences with X_n and Y_n defined on the same space for each n. Suppose $X_n \xrightarrow{D} X$ and $Y_n \xrightarrow{D} Y$, and assume X_n and Y_n are independent for all n and that X and Y are independent. Show that $X_n + Y_n \xrightarrow{D} X + Y$.

20 The Laws of Large Numbers

One of the fundamental results of Probability Theory is the Strong Law of Large Numbers. It helps to justify our intuitive notions of what probability actually is (Example 1), and it has many direct applications, such as (for example) Monte Carlo estimation theory (see Example 2).

Let $(X_n)_{n\geq 1}$ be a sequence of random variables defined on the same probability space and let $S_n = \sum_{j=1}^{n} X_j$. A theorem that states that $\frac{1}{n} S_n$ converges in some sense is a *law of large numbers*. There are many such results; for example L^2 ergodic theorems or the Birkhoff ergodic theorem, considered when the measure space is actually a probability space, are examples of laws of large numbers. (See Theorem 20.3, for example). The convergence can be in probability, in L^p, or almost sure. When the convergence is almost sure, we call it a *strong law of large numbers*.

Theorem 20.1 (Strong Law of Large Numbers). *Let $(X_n)_{n\geq 1}$ be independent and identically distributed (i.i.d.) and defined on the same space. Let*

$$\mu = E\{X_j\} \quad and \quad \sigma^2 = \sigma_{X_j}^2 < \infty.$$

Let $S_n = \sum_{j=1}^{n} X_j$. Then

$$\lim_{n\to\infty} \frac{S_n}{n} = \lim_{n\to\infty} \frac{1}{n} \sum_{j=1}^{n} X_j = \mu \ a.s. \ and \ in \ L^2.$$

Remark 20.1. We write μ, σ^2 instead of $\mu_j, \sigma_{X_j}^2$, since all the $(X_j)_{j\geq 1}$ have the same distribution and therefore the same mean and variance. Note also that $\lim_{n\to\infty} \frac{S_n}{n} = \mu$ in probability, since L^2 and a.s. convergence both imply convergence in probability. It is easy to prove $\lim_{n\to\infty} \frac{S_n}{n} = \mu$ in probability using Chebyshev's inequality, and this is often called the Weak Law of Large Numbers. Since it is a corollary of the Strong Law given here, we do not include its proof. The proof of Theorem 20.1 is also simpler if we assume only $X_j \in L^3$ (all j), and it is often presented this way in textbooks. A stronger result, where the X_n's are integrable but not necessarily square-integrable is stated in Theorem 20.2 and proved in Chapter 27.

Proof of Theorem 20.1: First let us note that without loss of generality we can assume $\mu = E\{X_j\} = 0$. Indeed if $\mu \neq 0$, then we can replace X_j with

$Z_j = X_j - \mu$. We obtain $\lim_{n\to\infty} \frac{1}{n}\sum_{j=1}^{n} Z_j = 0$ and therefore

$$\lim_{n\to\infty} \frac{1}{n}\sum_{j=1}^{n}(X_j - \mu) = \lim_{n\to\infty}\left(\frac{1}{n}\sum X_j\right) - \mu = 0$$

from which we deduce the result.

We henceforth assume $\mu = 0$. Recall $S_n = \sum_{j=1}^{n} X_j$ and let $Y_n = \frac{S_n}{n}$. Then $E\{Y_n\} = \frac{1}{n}\sum_{j=1}^{n} E\{X_j\} = 0$. Moreover $E\{Y_n^2\} = \frac{1}{n^2}\sum_{1\leq j,k\leq n} E\{X_j X_k\}$. However if $j \neq k$ then

$$E\{X_j X_k\} = E\{X_j\}E\{X_k\} = 0$$

since X_j and X_k are assumed to be independent. Therefore

$$E\{Y_n^2\} = \frac{1}{n^2}\sum_{j=1}^{n} E\{X_j^2\} \qquad (20.1)$$

$$= \frac{1}{n^2}\sum_{j=1}^{n} \sigma^2 = \frac{1}{n^2}(n\sigma^2)$$

$$= \frac{\sigma^2}{n}$$

and hence $\lim E\{Y_n^2\} = 0$.

Since Y_n converges to 0 in L^2 we know there is a subsequence converging to 0 a.s. However we want to conclude the *original* sequence converges a.s. To do this we find a subsequence converging a.s., and then treat the terms in between successive terms of the subsequence.

Since $E\{Y_n^2\} = \frac{\sigma^2}{n}$, let us choose the subsequence n^2; then

$$\sum_{n=1}^{\infty} E\{Y_{n^2}^2\} = \sum_{n=1}^{\infty} \frac{\sigma^2}{n^2} < \infty;$$

therefore by Theorem 9.2 we know $\sum_{n=1}^{\infty} Y_{n^2}^2 < \infty$ a.s., and hence the tail of this convergent series converges to 0; we conclude

$$\lim_{n\to\infty} Y_{n^2} = 0 \quad \text{a.s.} \qquad (20.2)$$

Next let $n \in \mathbf{N}$. Let $p(n)$ be the integer such that

$$p(n)^2 \leq n < (p(n) + 1)^2.$$

Then

$$Y_n - \frac{p(n)^2}{n}Y_{p(n)^2} = \frac{1}{n}\sum_{j=p(n)^2+1}^{n} X_j,$$

and as we saw in (20.1):

$$E\left\{\left(Y_n - \frac{p(n)^2}{n}Y_{p(n)^2}\right)^2\right\} = \frac{n - p(n)^2}{n^2}\sigma^2$$

$$\leq \frac{2p(n)+1}{n^2}\sigma^2,$$

$$\leq \frac{2\sqrt{n}+1}{n^2}\sigma^2 \leq \frac{3}{n^{\frac{3}{2}}}\sigma^2$$

because $p(n) \leq \sqrt{n}$.

Now we apply the same argument as before. We have

$$\sum_{n=1}^{\infty} E\left\{\left(Y_n - \frac{p(n)^2}{n}Y_{p(n)^2}\right)^2\right\} \leq \sum_{n=1}^{\infty} \frac{3\sigma^2}{n^{\frac{3}{2}}} < \infty.$$

Thus by Theorem 9.2 again, we have

$$\sum_{n=1}^{\infty}\left(Y_n - \frac{p(n)^2}{n}Y_{p(n)^2}\right)^2 < \infty \text{ a.s.}$$

which implies the tail converges to zero a.s. That is,

$$\lim_{n\to\infty}\left\{Y_n - \frac{p(n)^2}{n}Y_{p(n)^2}\right\} = 0 \text{ a.s.}$$

However since $\lim_{n\to\infty} Y_{p(n)^2} = 0$ a.s. by (20.2) and $\frac{p(n)^2}{n} \to 1$, we deduce $\lim_{n\to\infty} Y_n = 0$ a.s. as well. Recall $Y_n = \frac{S_n}{n}$, so the theorem is proved. $\square$

We give two other versions of Strong Laws of Large Numbers.

Theorem 20.2 (Kolmogorov's Strong Law of Large Numbers). *Let* (X_j) *be i.i.d. and* $\mu \in \mathbf{R}$. *Let* $S_n = \sum_{j=1}^{n} X_j$. *Then* $\lim_{n\to\infty} \frac{S_n}{n} = \mu$ *a.s. if and only if* $E\{X_j\} = \mu$. *In this case the convergence also holds in* L^1.

Remark 20.2. Note that Kolmogorov's strong law needs the (minimal) assumption that $(X_j)_{j\geq 1}$ are in L^1. An elegant way to prove Theorem 20.2 is to use the backwards martingale convergence theorem (see, e.g., Theorem 27.5).

Let $(\Omega, \mathcal{A}, P)$ be a probability space, and let $T : \Omega \to \Omega$ be one to one (i.e., injective) such that $T(\mathcal{A}) \subset \mathcal{A}$ (i.e., T maps measurable sets to measurable sets) and if $A \in \mathcal{A}$, then $P(T(A)) = P(A)$ (i.e., T *is measure preserving*). Let $T^2(\omega) = T(T(\omega))$ and define analogously powers of T. A set Λ is *invariant under* T if $1_\Lambda(\omega) = 1_\Lambda(T(\omega))$.

Theorem 20.3 (Ergodic Strong Law of Large Numbers). *Let T be a one-to-one measure preserving transformation of Ω onto itself. Assume the only T-invariant sets are sets of probability 0 or 1. If $X \in L^1$ then*

$$\lim_{n \to \infty} \frac{1}{n} \sum_{j=1}^{n} X(T^j(\omega)) = E\{X\}$$

a.s. and in L^1.

Theorem 20.3 is a consequence of the Birkhoff ergodic theorem; its advantage is that it replaces the hypothesis of independence with one of ergodicity. It is also called the *strong law of large numbers for stationary sequences of random variables.*

Example 1: In Example 17.1 we let $(X_j)_{j \geq 1}$ be a sequence of i.i.d. Bernoulli random variables, with $P(X_j = 1) = p$ and $P(X_j = 0) = q = 1 - p$ (all j). Then $S_n = \sum_{j=1}^{n} X_j$ is the number of "successes" in n trials, and $\frac{1}{n} S_n$ is the fraction of successes. The Strong Law of Large Numbers (Theorem 20.1) now tells us that

$$\lim_{n \to \infty} \frac{S_n}{n} = E\{X_1\} = p \text{ a.s.} \tag{20.3}$$

This gives, essentially, a justification to our claim that the probability of success is p. Thus in some sense this helps to justify the original axioms of probability we presented in Section 2, since we are finally able to deduce the intuitively pleasing result (20.3) from our original axioms.

Example 2: This is a simple example of a technique known as *Monte Carlo approximations.* (The etymology of the name is from the city of Monte Carlo of the Principality of Monaco, located in southern France. Gambling has long been legal there, and the name is a tribute to Monaco's celebration of the "laws of chance" through the operation of elegant gambling casinos.) Suppose f is a measurable function on $[0, 1]$, and $\int_0^1 |f(x)| dx < \infty$. Often we cannot obtain a closed form expression for $\alpha = \int_0^1 f(x) dx$ and we need to estimate it. If we let $(U_j)_{j \geq 1}$ be a sequence of independent uniform random variables on $[0, 1]$, and we call $I_n = \frac{1}{n} \sum_{j=1}^{n} f(U_j)$, then by Theorem 20.2 we have

$$\lim_{n \to \infty} \frac{1}{n} \sum_{j=1}^{n} f(U_j) = E\{f(U_j)\} = \int_0^1 f(x) dx,$$

a.s. and in L^2. Thus if we were to simulate the sequence $(U_j)_{j \geq 1}$ on a computer (using a random number generator to simulate uniform random variables, which is standard), we would get an approximation of $\int_0^1 f(x) dx$ for large n. This is just one method to estimate $\int_0^1 f(x) dx$, and it is usually not the best one except in the case where one wants to estimate a high dimensional integral: that is, if one wants to estimate $\int_{\mathbf{R}^d} f(x)dx$ for d large. The exact same ideas apply.

Example 3: ([7, p. 120]) Let Ω be a circle of radius $r = \frac{1}{2\pi}$. Let $\mathcal{A}$ be the Borel sets of the circle and let P be the Lebesgue measure on the circle (One can identify here the circle with the interval $[0, 1)$). Let α be irrational and T be rotation of Ω through α radians about the center of the circle. Then one can verify that T is injective, measure preserving, and that the invariant sets all have probability zero or one (this is where the irrationality of α comes in). Therefore by Theorem 20.3 we have

$$\lim_{n\to\infty} \frac{1}{n} \sum_{j=1}^{n} X(\omega + j\alpha) = \int_0^1 X(x)dx$$

for any $X \in L^1$ defined on Ω, for P-almost all x.

Exercises for Chapter 20

20.1 * (*A Weak Law of Large Numbers*). Let (X_j) be a sequence of random variables such that $\sup_j E\{X_j^2\} = c < \infty$ and $E\{X_j X_k\} = 0$ if $j \neq k$. Let $S_n = \sum_{j=1}^n X_j$.

a) Show that $P(|\frac{1}{n} S_n| \geq \varepsilon) \leq \frac{c}{n\varepsilon^2}$ for $\varepsilon > 0$;

b) $\lim_{n\to\infty} \frac{1}{n} S_n = 0$ in L^2 and in probability.

(*Note:* The usual i.i.d. assumptions have been considerably weakened here.)

20.2 Let $(Y_j)_{j\geq 1}$ be a sequence of independent Binomial random variables, all defined on the same probability space, and with law $B(p, 1)$. Let $X_n = \sum_{j=1}^n Y_j$. Show that X_j is $B(p, j)$ and that $\frac{X_j}{j}$ converges a.s. to p.

20.3 Let $(X_j)_{j\geq 1}$ be i.i.d. with X_j in L^1. Let $Y_j = e^{X_j}$. Show that

$$\left(\prod_{j=1}^n Y_j \right)^{\frac{1}{n}}$$

converges to a constant α a.s. [*Answer:* $\alpha = e^{E\{X_1\}}$.]

20.4 Let $(X_j)_{j\geq 1}$ be i.i.d. with X_j in L^1 and $E\{X_j\} = \mu$. Let $(Y_j)_{j\geq 1}$ be also i.i.d. with Y_j in L^1 and $E\{Y_j\} = \nu \neq 0$. Show that

$$\lim_{n\to\infty} \frac{1}{\sum_{j=1}^n Y_j} \sum_{j=1}^n X_j = \frac{\mu}{\nu} \quad \text{a.s.}$$

20.5 Let $(X_j)_{j\geq 1}$ be i.i.d. with X_j in L^1 and suppose $\frac{1}{\sqrt{n}} \sum_{j=1}^n (X_j - \nu)$ converges in distribution, to a random variable Z, Show that

$$\lim_{n\to\infty} \frac{1}{n} \sum_{j=1}^n X_j = \nu \quad \text{a.s.}$$

(*Hint:* If $Z_n = \frac{1}{\sqrt{n}} \sum_{j=1}^n (X_j - \nu)$, prove first that $\frac{1}{\sqrt{n}} Z_n$ converges in distribution to 0).

20.6 Let $(X_j)_{j\geq 1}$ be i.i.d. with X_j in L^p. Show that

$$\lim_{n\to\infty} \frac{1}{n} \sum_{j=1}^n X_j^p = E\{X^p\} \quad \text{a.s.}$$

20.7 Let $(X_j)_{j\geq 1}$ be i.i.d. $N(1, 3)$ random variables. Show that

$$\lim_{n\to\infty} \frac{X_1 + X_2 + \ldots + X_n}{X_1^2 + X_2^2 + \ldots + X_n^2} = \frac{1}{4} \quad \text{a.s.}$$

20.8 Let $(X_j)_{j \geq 1}$ be i.i.d. with mean μ and variance σ^2. Show that

$$\lim_{n \to \infty} \frac{1}{n} \sum_{i=1}^{n} (X_i - \mu)^2 = \sigma^2 \quad \text{a.s.}$$

20.9 Let $(X_j)_{j \geq 1}$ be i.i.d. integer valued random variables with $E\{|X_j|\} < \infty$. Let $S_n = \sum_{j=1}^{n} X_j$. $(S_n)_{n \geq 1}$ is called *a random walk on the integers*. Show that if $E(X_j) > 0$ then

$$\lim_{n \to \infty} S_n = \infty, \quad \text{a.s.}$$

21 The Central Limit Theorem

The Central Limit Theorem is one of the most impressive achievements of probability theory. From a simple description requiring minimal hypotheses, we are able to deduce precise results. The Central Limit Theorem thus serves as the basis for much of Statistical Theory. The idea is simple: let $X_1, \ldots, X_j, \ldots$ be a sequence of i.i.d. random variables with finite variance. Let $S_n = \sum_{j=1}^{n} X_j$. Then for n large, $\mathcal{L}(S_n) \approx N(n\mu, n\sigma^2)$, where $E\{X_j\} = \mu$ and $\sigma^2 = \text{Var}(X_j)$ (all j). The key observation is that *absolutely nothing* (except a finite variance) is assumed about the distribution of the random variables $(X_j)_{j\geq1}$. Therefore, if one can assume that a random variable in question is the sum of many i.i.d. random variables with finite variances, that one can infer that the random variable's distribution is approximately Gaussian. Next one can use data and do Statistical Tests to estimate μ and σ^2, and then one knows essentially everything!

Theorem 21.1 (Central Limit Theorem). *Let $(X_j)_{j\geq1}$ be i.i.d. with $E\{X_j\} = \mu$ and $Var(X_j) = \sigma^2$ (all j) with $0 < \sigma^2 < \infty$. Let $S_n = \sum_{j=1}^{n} X_j$. Let $Y_n = \frac{S_n - n\mu}{\sigma\sqrt{n}}$. Then Y_n converges in distribution to Y, where $\mathcal{L}(Y) = N(0, 1)$.*

Observe that if $\sigma^2 = 0$ above, then $X_j = \mu$ a.s. for all j, hence $\frac{S_n}{n} = \mu$ a.s.

Proof. Let φ_j be the characteristic function of $X_j - \mu$. Since the $(X_j)_{j\geq1}$ are i.i.d., φ_j does not depend on j and we write φ. Let $Y_n = \frac{S_n - n\mu}{\sigma\sqrt{n}}$. Since the X_j are independent, by Theorem 15.2

$$\varphi_{Y_n}(u) = \varphi_{\frac{1}{\sigma\sqrt{n}} \sum_{j=1}^{n}(X_j - \mu)}(u) \tag{21.1}$$

$$= \varphi_{\sum_{j=1}^{n}(X_j - \mu)}\left(\frac{u}{\sigma\sqrt{n}}\right)$$

$$= \prod_{j=1}^{n} \varphi_{(X_j - \mu)}\left(\frac{u}{\sigma\sqrt{n}}\right)$$

$$= \left(\varphi\left(\frac{u}{\sigma\sqrt{n}}\right)\right)^n.$$

Next note that $E\{X_j-\mu\} = 0$ and $E\{(X_j-\mu)^2\} = \sigma^2$, hence by Theorem 13.2 we know that φ has two continuous derivatives and moreover

$$\varphi'(u) = iE\left\{(X_j - \mu)e^{iu(X_j-\mu)}\right\},$$

$$\varphi''(u) = -E\left\{(X_j - \mu)^2 e^{iu(X_j-\mu)}\right\}.$$

Therefore $\varphi'(0) = 0$ and $\varphi''(0) = -\sigma^2$. If we expand φ in a Taylor expansion about $u = 0$, we get (see Exercise 14.4)

$$\varphi(u) = 1 + 0 - \frac{\sigma^2 u^2}{2} + u^2 h(u) \tag{21.2}$$

where $h(u) \to 0$ as $u \to 0$ (because φ'' is continuous). Recall from (21.1):

$$\varphi_{Y_n}(u) = \left(\varphi\left(\frac{u}{\sigma\sqrt{n}}\right)\right)^n$$

$$= e^{n\log\varphi(\frac{u}{\sigma\sqrt{n}})}$$

$$= e^{n\log(1-\frac{u^2}{2n}+\frac{u^2}{n\sigma^2}h(\frac{u}{\sigma\sqrt{n}}))},$$

where here "log" denotes the principal value of the complex valued logarithm. Taking limits as n tends to ∞ and using (for example) L'Hôpital's rule gives that

$$\lim_{n\to\infty} \varphi_{Y_n}(u) = e^{-u^2/2};$$

Lévy's Continuity Theorem (Theorem 19.1) then implies that Y_n converges in law to Z, where $\varphi_Z(u) = e^{-u^2/2}$; but then we know that $\mathcal{L}(Z) = N(0,1)$, using Example 13.5 and the fact that characteristic functions characterize distributions (Theorem 14.1). □

Let us now discuss the relationship between laws of large numbers and the central limit theorem. Let $(X_j)_{j\geq 1}$ be i.i.d. with finite variances, and let $\mu = E\{X_1\}$. Then by the Strong Law of Large Numbers,

$$\lim_{n\to\infty} \frac{S_n}{n} = \mu \quad \text{a.s. and in } L^2, \tag{21.3}$$

where $S_n = \sum_{j=1}^n X_j$. Thus we know the limit is μ, but a natural question is: How large must n be so that we are sufficiently close to μ? If we rewrite (21.3) as

$$\lim_{n\to\infty} \left|\frac{S_n}{n} - \mu\right| = 0 \quad \text{a.s. and in } L^2, \tag{21.4}$$

then what we wish to know is called a *rate of convergence*. We could ask, for example, does there exist an $\alpha \in \mathbf{R}$, $\alpha \neq 0$, such that

$$\lim_{n\to\infty} n^\alpha \left|\frac{S_n}{n} - \mu\right| = c \quad \text{a.s. } (c \neq 0)?$$

In fact, no such α exists. Indeed, one cannot have $n^\alpha(\frac{S_n}{n} - \mu)$ convergent to a non-zero constant or to a non-zero random variable a.s., or even in probability. However by the central limit theorem we know that if $\alpha = \frac{1}{2}$, $\sqrt{n}(\frac{S_n}{n} - \mu)$ converges in distribution to the normal distribution $N(0, \sigma^2)$. In this sense, the rate of convergence of the strong law of large numbers is $\sqrt{n}$.

One can weaken slightly the hypotheses of Theorem 21.1. Indeed with essentially the same proof, one can show:

Theorem 21.2 (Central Limit Theorem). *Let $(X_j)_{j\geq 1}$ be independent but not necessarily identically distributed. Let $E\{X_j\} = 0$ (all j), and let $\sigma_j^2 = \sigma_{X_j}^2$. Assume*

$$\sup_j E\{|X_j|^{2+\varepsilon}\} < \infty, \quad \text{some } \varepsilon > 0,$$

$$\sum_{j=1}^{\infty} \sigma_j^2 = \infty.$$

Then

$$\lim_{n\to\infty} \frac{S_n}{\sqrt{\sum_{j=1}^{n} \sigma_j^2}} = Z$$

where $\mathcal{L}(Z) = N(0,1)$ and where convergence is in distribution.

While Theorem 21.1 is, in some sense, the "classical" Central Limit Theorem, Theorem 21.2 shows it is possible to change the hypotheses and get similar results. As a consequence there are in fact many different central limit theorems, all similar in that they give sufficient conditions for properly normalized sums of random variables to converge in distribution to a normally distributed random variable. Indeed, martingale theory allows us to weaken the hypotheses of Theorem 21.2 substantially. See Theorem 27.7.

We note that one can also weaken the independence assumption to one of "asymptotic independence" via what is known as mixing conditions, but this is more difficult.

Finally, we note that Theorem 21.1 has a d-dimensional version which again has essentially the same proof.

Theorem 21.3 (Central Limit Theorem). *Let $(X_j)_{j\geq 1}$ be i.i.d. $\mathbf{R}^d$-valued random variables. Let the (vector) $\mu = E\{X_j\}$, and let Q denote the covariance matrix: $Q = (q_{k,\ell})_{1\leq k,\ell\leq d}$, where $q_{k,\ell} = \mathrm{Cov}(X_j^k, X_j^\ell)$, where X_j^k is the kth component of the $\mathbf{R}^d$-valued random variable X_j. Then*

$$\lim_{n\to\infty} \frac{S_n - n\mu}{\sqrt{n}} = Z$$

where $\mathcal{L}(Z) = N(0, Q)$ and where convergence is in distribution.

It is important to note that there is no requirement for the common covariance matrix Q to be invertible in Theorem 21.3. In this way we see that the Central Limit Theorem gives rise to gaussian limits without densities (since an $\mathbf{R}^d$ valued Gaussian r.v. has a density if and only if its covariance matrix Q is invertible).

Examples:

1. Let $(X_j)_{j\geq 1}$ be i.i.d. with $P(X_j = 1) = p$ and $P(X_j = 0) = q = 1 - p$. Then $S_n = \sum_{j=1}^{n} X_j$ is Binomial $(\mathcal{L}(S_n) = B(p, n))$. We have $\mu = E\{X_j\} = p$ and $\sigma^2 = \sigma^2_{X_j} = pq = p(1 - p)$. By the Strong Law of Large Numbers we have

$$\lim_{n\to\infty} \frac{S_n}{n} = p \text{ a.s.}$$

and by the Central Limit Theorem (Theorem 21.1) we have (with convergence being in distribution);

$$\frac{S_n - np}{\sqrt{np(1 - p)}} \xrightarrow{D} Z$$

where $\mathcal{L}(Z) = N(0, 1)$.

2. Suppose $(X_j)_{j\geq 1}$ are i.i.d. random variables, all in L^2, and with (common) distribution function F. We assume F is unknown and we would like to estimate it. We give here a standard technique to do just that. Let

$$Y_j(x) = 1_{\{X_j \leq x\}}.$$

Note that Y_j are i.i.d. and in L^2. Next define

$$F_n(x) = \frac{1}{n} \sum_{j=1}^{n} Y_j(x), \text{ for } x \text{ fixed.}$$

The function $F_n(x)$ defined on $\mathbf{R}$ is called the *empirical distribution function* (it should indeed be written as $F_n(x, \omega)$, since it depends on ω!). By the Strong Law of Large numbers we have

$$\lim_{n\to\infty} F_n(x) = \lim_{n\to\infty} \frac{1}{n} \sum_{j=1}^{n} Y_j(x) = E\{Y_1(x)\}.$$

However,

$$E\{Y_1(x)\} = E\{1_{\{X_j \leq x\}}\} = P(X_j \leq x) = F(x),$$

and thus we can conclude

$$\lim_{n\to\infty} F_n(x) = F(x) \text{ a.s.}$$

That is, *the empirical distribution function converges to the actual distribution function a.s. and in L^2*. With a little more work we can obtain a stronger result: $\lim_{n\to\infty} \sup_x |F_n(x) - F(x)| = 0$ a.s.. This is known as the *Glivenko–Cantelli Theorem*. Using the Central Limit Theorem we can moreover show that the rate of convergence is $\sqrt{n}$: indeed, since $F_n(x) - F(x)$ tends to 0, we can hope to find a rate by showing $n^\alpha(F_n(x) - F(x))$ converges for some α. But

$$\sqrt{n}(F_n(x) - F(x)) = \sqrt{n}\left(\frac{1}{n}\sum_{j=1}^n Y_j(x) - E\{Y_1(x)\}\right)$$

$$= \frac{\sum_{j=1}^n Y_j(x) - nE\{Y_1(x)\}}{\sqrt{n}}$$

and hence by Theorem 21.1 it converges to a normal random variable Z with $\mathcal{L}(Z) = N(0, \sigma^2(x))$, and where $\sigma^2(x) = \text{Var}(Y_1(x)) = F(x)(1 - F(x))$.

Example 2 raises an interesting question: how large must n be before the empirical distribution function is "close" to the actual distribution function? In essence this is equivalent to asking for a rate of convergence result for the Central Limit Theorem. (Recall that we have already seen that the Central Limit Theorem itself gives a rate of convergence of $\sqrt{n}$ for the Strong Law of Large Numbers.) A classic result is the following:

Theorem 21.4 (Berry-Esseen). *Let* $(X_j)_{j\geq 1}$ *be i.i.d. and suppose* $E\{|X_j|^3\} < \infty$. *Let* $G_n(x) = P(\frac{S_n - n\mu}{\sigma\sqrt{n}} \leq x)$ *where* $\mu = E\{X_j\}$ *and* $\sigma^2 = \sigma^2_{X_j} < \infty$. *Let* $\Phi(x) = P(Z \leq x)$, *where* $\mathcal{L}(Z) = N(0, 1)$. *Then*

$$\sup_x |G_n(x) - \Phi(x)| \leq c\frac{E\{|X_1|^3\}}{\sigma^3\sqrt{n}}$$

for a constant c.

The proof of Theorem 21.4 is too advanced for this book. The interested reader can consult [8, p.108] where it is proved for $c = 3$. (The current best estimates are $c = 0.7975$, see [24], and I.S. Shiganov has shown $c \leq 0.7655$. Also Esseen established a lower bound. Thus current knowledge is $0.4097 \leq c \leq 0.7655$.)

Exercises for Chapter 21

21.1 Let $(X_j)_{j\geq 1}$ be i.i.d. with $P(X_j = 1) = P(X_j = 0) = \frac{1}{2}$. Let $S_n = \sum_{j=1}^{n} X_j$, and let $Z_n = 2S_n - n$. (Z_n is the excess of heads over tails in n tosses, if $X_j = 1$ when heads and $X_j = 0$ when tails on the jth toss.) Show that

$$\lim_{n\to\infty} P\left(\frac{Z_n}{\sqrt{n}} < x\right) = \Phi(x)$$

where

$$\Phi(x) = \frac{1}{\sqrt{2\pi}} \int_{-\infty}^{x} e^{-u^2/2} du.$$

21.2 Let $(X_j)_{j\geq 1}$ be independent, double exponential with parameter 1 (that is, the common density is $\frac{1}{2}e^{-|x|}$, $-\infty < x < \infty$). Show that

$$\lim_{n\to\infty} \sqrt{n}\left(\frac{\sum_{j=1}^{n} X_j}{\sum_{j=1}^{n} X_j^2}\right) = Z,$$

where $\mathcal{L}(Z) = N(0, \frac{1}{2})$, and where convergence is in distribution. (*Hint:* Use Slutsky's theorem (Exercise 18.14).)

21.3 Construct a sequence of random variables $(X_j)_{j\geq 1}$, independent, such that $\lim_{j\to\infty} X_j = 1$ in probability, and $E\{X_j^2\} \geq j$. Let Y be independent of the sequence $(X_j)_{j\geq 1}$, and $\mathcal{L}(Y) = N(0, 1)$. Let $Z_j = YX_j$, $j \geq 1$. Show that

a) $E\{Z_j\} = 0$
b) $\lim_{j\to\infty} \sigma_{Z_j}^2 = \infty$
c) $\lim_{j\to\infty} Z_j = Z$ in distribution, where $\mathcal{L}(Z) = N(0, 1)$.

(*Hint:* To construct X_j, let $(\Omega_j, \mathcal{A}_j, P_j)$ be $([0, 1], \mathcal{B}[0, 1], m(ds))$, where m is Lebesgue measure on $[0, 1]$. Let

$$X_j(\omega) = (j + 1)1_{[0,1/j]}(\omega) + 1_{(1/j,1]}(\omega),$$

and take the infinite product as in Theorem 10.4. To prove (c) use Slutsky's theorem (Exercise 18.14)). (Note that the hypotheses of the central limit theorems presented here are not satisfied; of course, the theorems give sufficient conditions, not necessary ones.)

21.4 (Durrett, [8]). Let $(X_j)_{j\geq 1}$ be i.i.d. nonnegative with $E\{X_1\} = 1$ and $\sigma_{X_1}^2 = \sigma^2 \in (0, \infty)$. Show that

$$\frac{2}{\sigma}(\sqrt{S_n} - \sqrt{n}) \xrightarrow{\mathcal{D}} Z,$$

with $\mathcal{L}(Z) = N(0, 1)$.

$$\left(\textit{Hint:} \quad \frac{S_n - n}{\sqrt{n}} = \frac{(\sqrt{S_n} + \sqrt{n})}{\sqrt{n}}(\sqrt{S_n} - \sqrt{n}).\right)$$

21.5 Let (X_j) be i.i.d. Poisson random variables with parameter $\lambda = 1$. Let $S_n = \sum_{j=1}^n X_j$. Show that $\lim_{n\to\infty} \frac{S_n - n}{\sqrt{n}} = Z$, where $\mathcal{L}(Z) = N(0,1)$.

21.6 Let Y^λ be a Poisson random variable with parameter $\lambda > 0$. Show that

$$\lim_{\lambda\to\infty} \frac{Y^\lambda - \lambda}{\sqrt{\lambda}} = Z$$

where $\mathcal{L}(Z) = N(0,1)$ and convergence is in distribution. (*Hint:* Use Exercise 21.5 and compare Y^λ with $S_{[\lambda]}$ and $S_{[\lambda]+1}$, where $[\lambda]$ denotes the largest integer less than or equal to λ.)

21.7 Show that

$$\lim_{n\to\infty} e^{-n} \left(\sum_{k=0}^n \frac{n^k}{k!} \right) = \frac{1}{2}.$$

(*Hint:* Use Exercise 21.5.)

21.8 Let $(X_j)_{j\geq 1}$ be i.i.d. with $E\{X_j\} = 0$ and $\sigma_{X_j}^2 = \sigma^2 < \infty$. Let $S_n = \sum_{i=1}^n X_i$. Show that $\frac{S_n}{\sigma\sqrt{n}}$ does not converge in probability.

21.9 * Let $(X_j)_{j\geq 1}$ be i.i.d. with $E\{X_j\} = 0$ and $\sigma_{X_j}^2 = \sigma^2 < \infty$. Let $S_n = \sum_{j=1}^n X_j$. Show that

$$\lim_{n\to\infty} E\left\{ \frac{|S_n|}{\sqrt{n}} \right\} = \sqrt{\frac{2}{\pi}}\sigma.$$

(*Hint:* Let $\mathcal{L}(Z) = N(0,\sigma^2)$ and calculate $E\{|Z|\}$.)

21.10 (Gut, [11]). Let $(X_j)_{j\geq 1}$ be i.i.d. with the uniform distribution on $(-1,1)$. Let

$$Y_n = \frac{\sum_{j=1}^n X_j}{\sum_{j=1}^n X_j^2 + \sum_{j=1}^n X_j^3}.$$

Show that $\sqrt{n}Y_n$ converges. (*Answer:* $\sqrt{n}Y_n$ converges in distribution to Z where $\mathcal{L}(Z) = N(0,3)$.)

21.11 Let $(X_j)_{j\geq 1}$ be independent and let X_j have the uniform distribution on $(-j,j)$.

a) Show that

$$\lim_{n\to\infty} \frac{S_n}{n^{\frac{3}{2}}} = Z$$

in distribution where $\mathcal{L}(Z) = N(0, \frac{1}{9})$ (*Hint:* Show that the characteristic function of X_j is $\varphi_{X_j}(u) = \frac{\sin(uj)}{uj}$; compute $\varphi_{S_n}(u)$, then $\varphi_{S_n/n^{3/2}}(u)$, and prove that the limit is $e^{-u^2/18}$ by using $\sum_{j=1}^n j^2 = \frac{n(n+1)(2n+1)}{6}$).

b) Show that

$$\lim_{n\to\infty} \frac{S_n}{\sqrt{\sum_{j=1}^{n} \sigma_j^2}} = Z$$

in distribution, where $\mathcal{L}(Z) = N(0,1)$. (*Note:* This is *not* a particular case of Theorem 21.2).

21.12 * Let $X \in L^2$ and suppose X has the same distribution as $\frac{1}{\sqrt{2}}(Y + Z)$, where Y, Z are independent and X, Y, Z all have the same distribution. Show that X is $N(0, \sigma^2)$ with $\sigma^2 < \infty$. (*Hint:* Show by iteration that X has the same law as $\frac{1}{\sqrt{n}} \sum_{i=1}^{n} X_i$ with (X_i) i.i.d., for $n = 2^m$.)

22 L^2 and Hilbert Spaces

We suppose given a probability space $(\Omega, \mathcal{F}, P)$. Let L^2 denote all (equivalence classes for a.s. equality of) random variables X such that $E\{X^2\} < \infty$. We henceforth identify all random variables X, Y in L^2 that are equal a.s. and consider them to be representatives of the same random variable. This has the consequence that if $E\{X^2\} = 0$, we can conclude that $X = 0$ (and not only $X = 0$ a.s.).

We can define an inner product in L^2 as follows: for X, Y in L^2, define

$$\langle X, Y \rangle = E\{XY\}.$$

Note that $|E\{XY\}| \leq E\{X^2\}^{\frac{1}{2}} E\{Y^2\}^{\frac{1}{2}} < \infty$ by the Cauchy–Schwarz inequality. We have seen in Theorem 9.3 that L^2 is a *linear space:* if X, Y are both in L^2, and α, β are constants, then $\alpha X + \beta Y$ is in L^2 as well. We further note that the inner product is linear in each component: For example

$$\langle \alpha X + \beta Y, Z \rangle = \alpha \langle X, Z \rangle + \beta \langle Y, Z \rangle.$$

Finally, observe that

$$\langle X, X \rangle \geq 0, \text{ and } \langle X, X \rangle = 0 \text{ if and only if } X = 0 \text{ a.s.}$$

since $X = 0$ a.s. implies $X = 0$ by our convention of identifying almost surely equal random variables. This leads us to define a norm for L^2 as follows:

$$\|X\| = \langle X, X \rangle^{\frac{1}{2}} = (E\{X^2\})^{\frac{1}{2}}.$$

We then have $\|X\| = 0$ implies $X = 0$ (recall that in L^2, $X = 0$ is the same as $X = 0$ a.s.), and by bilinearity and the Cauchy–Schwarz inequality we get

$$\begin{aligned} \|X + Y\|^2 &= E\{X^2\} + 2E\{XY\} + E\{Y^2\} \\ &\leq \|X\|^2 + 2\|X\| \, \|Y\| + \|Y\|^2 \\ &= (\|X\| + \|Y\|)^2, \end{aligned}$$

and thus we obtain Minkowski's inequality:

$$\|X + Y\| \leq \|X\| + \|Y\|,$$

so that our norm satisfies the triangle inequality and is a true norm. We have shown the following:

Theorem 22.1. L^2 *is a normed linear space with an inner product* $\langle \cdot, \cdot \rangle$. *Moreover one has* $\| \cdot \| = \langle \cdot, \cdot \rangle^{\frac{1}{2}}$.

We next want to show that L^2 is a *complete* normed linear space; that is, if X_n is a sequence of random variables that is Cauchy under $\| \cdot \|$, then there exists a limit in L^2 (recall that X_n is *Cauchy* if $\|X_n - X_m\| \to 0$ when both m and n tend to infinity; every convergent sequence is Cauchy). Theorem 22.2 is sometimes known as the *Riesz–Fischer Theorem*.

Theorem 22.2. L^2 *is complete.*

Proof. Let X_n be a Cauchy sequence in L^2. That is, for any $\varepsilon > 0$, there exists N such that $n, m \geq N$ implies $\|X_n - X_m\| \leq \varepsilon$. Choose a sequence of epsilons of the form $\frac{1}{2^n}$. Then we have a subsequence $(X_{n_k})_{k \geq 1}$ such that $\|X_{n_k} - X_{n_{k+1}}\| \leq \frac{1}{2^k}$.

Define
$$Y_n = \sum_{p=1}^{n} |X_{n_p} - X_{n_{p+1}}|.$$

By the triangle inequality we have

$$E\{Y_n^2\} \leq \left(\sum_{p=1}^{n} \|X_{n_p} - X_{n_{p+1}}\| \right)^2 \leq 1.$$

Let $Y = \lim_{n \to \infty} Y_n$ which exists because $Y_n(\omega)$ is a nondecreasing sequence, for each ω (a.s.). Since $E\{Y_n^2\} \leq 1$ each n, by the Monotone Convergence Theorem (Theorem 9.1(d)) $E\{Y^2\} \leq 1$ as well. Therefore $Y < \infty$ a.s., and hence the sequence $X_{n_1} + \sum_{p=1}^{\infty} (X_{n_{p+1}} - X_{n_p})$ converges absolutely a.s. Since it is a telescoping series we conclude $X_{n_p}(\omega)$ converges toward a limit $X(\omega)$ as $p \to \infty$, and moreover $|X(\omega)| \leq |X_{n_1}(\omega)| + Y(\omega)$. Since X_{n_1} and Y are in L^2, so also $X \in L^2$.

Next, note that

$$X - X_{n_p} = \lim_{m \to \infty} Z_m^p = \lim_{m \to \infty} \sum_{q=p}^{m} (X_{n_{q+1}} - X_{n_q}).$$

Since $|Z_m^p| \leq Y$ for each p, m, by Lebesgue's dominated convergence theorem (Theorem 9.1(f)) we have

$$\|X - X_{n_p}\| = \lim_{m \to \infty} \|Z_m^p\| \leq \lim_{m} \sum_{q=p}^{m} \|X_{n_{q+1}} - X_{n_q}\| \leq \frac{1}{2^{p-1}}$$

and we conclude $\lim_{p \to \infty} \|X - X_{n_p}\| = 0$. Therefore X_{n_p} converges to X in L^2.

Finally,

$$\|X_n - X\| \le \|X_n - X_{n_p}\| + \|X_{n_p} - X\|.$$

Hence letting n and p go to infinity, we deduce that X_n tends to X in L^2.

□

Definition 22.1. *A Hilbert space $\mathcal{H}$ is a complete normed linear space with an inner product satisfying $\langle x, x \rangle^{\frac{1}{2}} = \|x\|$, all $x \in \mathcal{H}$.*

We now have established:

Theorem 22.3. *L^2 is a Hilbert space.*

Henceforth we will describe results for Hilbert spaces; of course these results apply as well for L^2. *From now on $\mathcal{H}$ will denote a Hilbert space* with norm $\|\cdot\|$ and inner product $\langle\cdot,\cdot\rangle$, while α an β below always denote real numbers.

Definition 22.2. *Two vectors x and y in $\mathcal{H}$ are* orthogonal *if $\langle x, y \rangle = 0$. A vector x is orthogonal to a set of vectors Γ if $\langle x, y \rangle = 0$ for every $y \in \Gamma$.*

Observe that if $\langle x, y \rangle = 0$ then $\|x + y\|^2 = \|x\|^2 + \|y\|^2$; this is a Hilbert space version of the Pythagorean theorem.

Theorem 22.4 (Continuity of the inner product). *If $x_n \to x$ and $y_n \to y$ in $\mathcal{H}$, then $\langle x_n, y_n \rangle \to \langle x, y \rangle$ in $\mathbf{R}$* (and thus also $\|x_n\| \to \|x\|$).

Proof. The Cauchy–Schwarz inequality implies $\langle x, y \rangle \le \|x\|\,\|y\|$, hence

$$|\langle x, y \rangle - \langle x_n, y_n \rangle| = |\langle x - x_n, y_n \rangle + \langle x - x_n, y - y_n \rangle + \langle x_n, y - y_n \rangle|$$
$$\le \|x - x_n\|\,\|y_n\| + \|x - x_n\|\,\|y - y_n\| + \|x_n\|\,\|y - y_n\|.$$

Note that $\sup_n \|y_n\| < \infty$ and $\sup_n \|x_n\| < \infty$, since x_n and y_n are both convergent sequences in $\mathcal{H}$ (for example, $\|x_n\| \le \|x_n - x\| + \|x\|$ and $\|x\| < \infty$ and $\|x_n - x\| \to 0$). Thus the right side of the above inequality tends to 0 as n tends to ∞.

□

Definition 22.3. *A subset $\mathcal{L}$ of $\mathcal{H}$ is called a* subspace *if it is linear (that is, $x, y \in \mathcal{L}$ implies $\alpha x + \beta y \in \mathcal{L}$) and if it is closed (that is, if $(x_n)_{n\ge 1}$ converges to x in $\mathcal{H}$, then $x \in \mathcal{L}$).*

Theorem 22.5. *Let Γ be a set of vectors. Let $\Gamma^\perp$ denote all vectors orthogonal to all vectors in Γ. Then $\Gamma^\perp$ is a subspace of $\mathcal{H}$.*

Proof. First note that $\Gamma^\perp$ is a linear space, even if Γ is not. Indeed, if $x, y \in \Gamma^\perp$, then $\langle x, z \rangle = 0$ and $\langle y, z \rangle = 0$, for each $z \in \Gamma$. Therefore

$$\langle \alpha x + \beta y, z \rangle = \alpha \langle x, z \rangle + \beta \langle y, z \rangle = 0,$$

and $\alpha x + \beta y \in \Gamma^\perp$ also. It follows from Theorem 22.4 that $\Gamma^\perp$ is closed. □

Definition 22.4. *For a subspace $\mathcal{L}$ of $\mathcal{H}$, let $d(x,\mathcal{L}) = \inf\{\|x - y\|; y \in \mathcal{L}\}$ denote the distance from $x \in \mathcal{H}$ to $\mathcal{L}$.*

Note that if $\mathcal{L}$ is a subspace, then $x \in \mathcal{L}$ iff $d(x,\mathcal{L}) = 0$ (recall that a linear subspace of a closed space is always closed).

Theorem 22.6. *Let $\mathcal{L}$ be a subspace of $\mathcal{H}$; $x \in \mathcal{H}$. There is a unique vector $y \in \mathcal{L}$ such that $\|x - y\| = d(x,\mathcal{L})$.*

Proof. If $x \in \mathcal{L}$, then $y = x$. If x is not in $\mathcal{L}$, let $y_n \in \mathcal{L}$ such that $\lim_{n\to\infty} \|x - y_n\| = d(x,\mathcal{L})$. We want to show that $(y_n)_{n \geq 1}$ is Cauchy in $\mathcal{H}$. Note first that

$$\|y_n - y_m\|^2 = \|x - y_m\|^2 + \|x - y_n\|^2 - 2\langle x - y_m, x - y_n\rangle. \tag{22.1}$$

We use the inequality

$$\left\| x - \frac{y_m + y_n}{2} \right\| \leq \frac{\|x - y_m\|}{2} + \frac{\|x - y_n\|}{2}$$

to conclude that

$$\lim_{n,m\to\infty} \left\| x - \frac{y_m + y_n}{2} \right\| \leq d(x,\mathcal{L}),$$

hence

$$\lim_{n,m\to\infty} \left\| x - \frac{y_m + y_n}{2} \right\| = d(x,\mathcal{L}),$$

since $d(x,\mathcal{L})$ is an infimum and $\frac{y_m + y_n}{2} \in \mathcal{L}$ because $\mathcal{L}$ is a subspace. We now have

$$d(x,\mathcal{L})^2 = \lim_{m,n\to\infty} \left\| x - \frac{y_m + y_n}{2} \right\|^2$$
$$= \lim_{m,n\to\infty} \left\{ \|x - y_m\|^2 + \|x - y_n\|^2 + 2\langle x - y_m, x - y_n\rangle \right\}/4$$

and therefore

$$\lim_{n,m\to\infty} \langle x - y_m, x - y_n\rangle = d(x,\mathcal{L})^2. \tag{22.2}$$

If we now combine (22.1) and (22.2) we see that $(y_n)_{n \geq 1}$ is Cauchy. Therefore $\lim y_n = y$ exists and is in $\mathcal{L}$, since $\mathcal{L}$ is closed. Moreover $d(x,\mathcal{L}) = \|x - y\|$, by the continuity of the distance function.

It remains to show the uniqueness of y. Suppose z were another such vector in $\mathcal{L}$. Then the sequence

$$w_{2n} = y,$$
$$w_{2n+1} = z,$$

is again a Cauchy sequence in $\mathcal{L}$ by the previous argument, and hence it converges to a unique limit; whence $y = z$. $\qquad\square$

We now consider the important concept of *projections*. We fix our Hilbert space $\mathcal{H}$ and our (closed, linear) subspace $\mathcal{L}$. The *projection of a vector x in $\mathcal{H}$ onto $\mathcal{L}$* consists of taking the (unique) $y \in \mathcal{L}$ which is closest to x. We let Π denote this projection operator. The next theorem gives useful properties of Π.

Theorem 22.7. *The projection operator Π of $\mathcal{H}$ onto a subspace $\mathcal{L}$ satisfies the following three properties:*

(i) *Π is idempotent: $\Pi^2 = \Pi$;*
(ii) *$\Pi x = x$ for $x \in \mathcal{L}$; $\Pi x = 0$ for $x \in \mathcal{L}^\perp$;*
(iii) *For every $x \in \mathcal{H}$, $x - \Pi x$ is orthogonal to $\mathcal{L}$.*

Proof. (i) follows immediately from the definition of projection.

(ii) If $x \in \mathcal{L}$, then $d(x, \mathcal{L}) = 0$, and since x is closest to x ($\|x - x\| = 0$), $\Pi x = x$. Moreover if $x \in \mathcal{L}^\perp$, then $\|x - y\|^2 = \langle x - y, x - y \rangle = \|x\|^2 + \|y\|^2$ for $y \in \mathcal{L}$, and thus $y = 0$ minimizes $d(x, \mathcal{L})$; hence $\Pi x = 0$.

(iii) We first note that, for $y \in \mathcal{L}$:

$$\|x - \Pi x\|^2 \leq \|x - (\Pi x + y)\|^2$$
$$= \|x - \Pi x\|^2 + \|y\|^2 - 2\langle x - \Pi x, y \rangle,$$

and therefore

$$2\langle x - \Pi x, y \rangle \leq \|y\|^2.$$

Since $y \in \mathcal{L}$ was arbitrary and since $\mathcal{L}$ is linear we can replace y with αy, any $\alpha \in \mathbf{R}_+$, to obtain

$$2\langle x - \Pi x, \alpha y \rangle \leq \|\alpha y\|^2,$$

and dividing by α gives

$$2\langle x - \Pi x, y \rangle \leq \alpha \|y\|^2;$$

we let α tend to zero to conclude $\langle x - \Pi x, y \rangle \leq 0$. Analogously we obtain $\langle x - \Pi x, y \rangle \geq 0$ by considering negative α. Thus $x - \Pi x$ is orthogonal to $\mathcal{L}$.

$\square$

Corollary 22.1. *Let Π be the projection operator of $\mathcal{H}$ onto a subspace $\mathcal{L}$. Then $x = \Pi x + (x - \Pi x)$ is a unique representation of x as the sum of a vector in $\mathcal{L}$ and one in $\mathcal{L}^\perp$. Such a representation exists. Moreover $x - \Pi x$ is the projection of x onto $\mathcal{L}^\perp$; and $(\mathcal{L}^\perp)^\perp = \mathcal{L}$.*

Proof. The existence of such a representation is shown in Theorem 22.7(iii). As for uniqueness, let $x = y + z$ be another such representation. Then $y - \Pi x = z - (x - \Pi x)$ is a vector simultaneously in $\mathcal{L}$ and $\mathcal{L}^{\perp}$; therefore it must be 0 (because it is orthogonal to itself), and we have uniqueness.

Next observe that $\mathcal{L} \subset (\mathcal{L}^{\perp})^{\perp}$. Indeed, if $x \in \mathcal{L}$ and $y \in \mathcal{L}^{\perp}$ then $\langle x, y \rangle = 0$, so $x \in (\mathcal{L}^{\perp})^{\perp}$. On the other hand if $x \in (\mathcal{L}^{\perp})^{\perp}$, then $x = y + z$ with $y \in \mathcal{L}$ and $z \in \mathcal{L}^{\perp}$. But z must be 0, since otherwise we have $\langle x, z \rangle = \langle y, z \rangle + \langle z, z \rangle$, and $\langle y, z \rangle = 0$ since $y \in \mathcal{L}$ and $z \in \mathcal{L}^{\perp}$; and also $\langle x, z \rangle = 0$ since $z \in \mathcal{L}^{\perp}$ and $x \in (\mathcal{L}^{\perp})^{\perp}$. Thus $\langle z, z \rangle = 0$, hence $z = 0$. Therefore $x = y$, with $y \in \mathcal{L}$, hence $x \in \mathcal{L}$, and $(\mathcal{L}^{\perp})^{\perp} \subset \mathcal{L}$. □

Corollary 22.2. *Let Π be the projection operator $\mathcal{H}$ onto a subspace $\mathcal{L}$. Then*

(i) $\langle \Pi x, y \rangle = \langle x, \Pi y \rangle$,
(ii) Π *is a linear operator:* $\Pi(\alpha x + \beta y) = \alpha \Pi x + \beta \Pi y$.

Proof. (i) By Corollary 22.1 we write uniquely:

$$x = x_1 + x_2, \qquad x_1 \in \mathcal{L}; x_2 \in \mathcal{L}^{\perp},$$
$$y = y_1 + y_2, \qquad y_1 \in \mathcal{L}; y_2 \in \mathcal{L}^{\perp}.$$

Then

$$\langle \Pi x, y \rangle = \langle x_1, y \rangle = \langle x_1, y_1 + y_2 \rangle = \langle x_1, y_1 \rangle,$$

since $\langle x_1, y_2 \rangle = 0$. Continuing in reverse for y, and using $\langle x_2, y_1 \rangle = 0$:

$$= \langle x_1 + x_2, y_1 \rangle = \langle x, y_1 \rangle = \langle x, \Pi y \rangle.$$

(ii) Again using the unique decomposition of Corollary 22.1, we have:

$$\alpha x + \beta y = (\alpha x_1 + \beta y_1) + (\alpha x_2 + \beta y_2),$$

hence

$$\Pi(\alpha x + \beta y) = \alpha x_1 + \beta y_1 = \alpha \Pi x + \beta \Pi y.$$

□

We end this treatment with a converse that says, in essence, that if an operator behaves like a projection then it is a projection.

Theorem 22.8. *Let T map $\mathcal{H}$ onto a subspace $\mathcal{L}$. Suppose that $x - Tx$ is orthogonal to $\mathcal{L}$ for all $x \in \mathcal{H}$. Then $T = \Pi$, the projection operator onto the subspace $\mathcal{L}$.*

Proof. We can write $x = Tx + (x - Tx)$, with $Tx \in \mathcal{L}$ and $(x - Tx) \in \mathcal{L}^{\perp}$. By Corollary 22.1 to Theorem 22.7, Tx must be the projection of x onto $\mathcal{L}$. □

Exercises for Chapter 22

22.1 Using that $(a - b)^2 \geq 0$, prove that $(a + b)^2 \leq 2a^2 + 2b^2$.

22.2 Let $x, y \in \mathcal{H}$, a Hilbert space, with $\langle x, y \rangle = 0$. Prove the Pythagorean Theorem: $\|x + y\|^2 = \|x\|^2 + \|y\|^2$.

22.3 Show that $\mathbf{R}^n$ is a Hilbert space with an inner product given by the "dot product" if $\vec{x} = (x_1, \ldots, x_n)$ and $\vec{y} = (y_1, \ldots, y_n)$, then $\langle \vec{x}, \vec{y} \rangle = \sum_{i=1}^{n} x_i y_i$.

22.4 Let $\mathcal{L}$ be a linear subspace of $\mathcal{H}$ and Π projection onto $\mathcal{L}$. Show that Πy is the unique element of $\mathcal{L}$ such that $\langle \Pi y, z \rangle = \langle y, z \rangle$, for all $z \in \mathcal{L}$.

23 Conditional Expectation

Let X and Y be two random variables with Y taking values in $\mathbf{R}$ with X taking on *only countably many values*. It often arises that we know already the value of X and want to calculate the expected value of Y taking into account the knowledge of X. That is, suppose we know that the event $\{X = j\}$ for some value j has occurred. The expectation of Y may change given this knowledge. Indeed, if $Q(\Lambda) = P(\Lambda|X = j)$, it makes more sense to calculate $E_Q\{Y\}$ than it does to calculate $E_P\{Y\}$ ($E_R\{\cdot\}$ denotes expectation with respect to the Probability measure R.)

Definition 23.1. *Let X have values $\{x_1, x_2, \ldots, x_n, \ldots\}$ and Y be a random variable. Then if $P(X = x_j) > 0$ the conditional expectation of Y given $\{X = x_j\}$ is defined to be*

$$E\{Y|X = x_j\} = E_Q\{Y\},$$

where Q is the probability given by $Q(\Lambda) = P(\Lambda|X = x_j)$, provided $E_Q\{|Y|\} < \infty$.

Theorem 23.1. *In the previous setting, and if further Y is countably valued with values $\{y_1, y_2, \ldots, y_n, \ldots\}$ and if $P(X = x_j) > 0$, then*

$$E\{Y|X = x_j\} = \sum_{k=1}^{\infty} y_k P(Y = y_k | X = x_j),$$

provided the series is absolutely convergent.

Proof.

$$E\{Y|X = x_j\} = E_Q\{Y\} = \sum_{k=1}^{\infty} y_k Q(Y = y_k) = \sum_{k=1}^{\infty} y_k P(Y = y_k | X = x_j).$$

$\square$

Next, still with X having at most a countable number of values, we wish to define the conditional expectation of any real valued r.v. Y given knowledge of the *random variable* X, rather than given only the event $\{X = x_j\}$. To this effect we consider the function

$$f(x) = \begin{cases} E\{Y|X = x\} & \text{if } P(X = x) > 0 \\ \text{any arbitrary value} & \text{if } P(X = x) = 0. \end{cases} \tag{23.1}$$

Definition 23.2. *Let X be countably valued and let Y be a real valued random variable. The conditional expectation of Y given X is defined to be*

$$E\{Y|X\} = f(X),$$

where f is given by (23.1) provided f is well defined (that is, Y is integrable with respect to the probability measure Q_j defined by by $Q_j(\Lambda) = P(\Lambda|X = x_j)$, for all j such that $P(X = x_j) > 0$).

Remark 23.1. The above definition does not really define $E\{Y|X\}$ everywhere, but only almost everywhere since it is arbitrary on each set $\{X = x\}$ such that $P(X = x) = 0$: this will be a distinctive feature of the conditional expectation for more general r.v. X's as defined below.

Example: Let X be a Poisson random variable with parameter λ. When $X = n$, we have that each one of the n outcomes has a probability of success p, independently of the others. Let S denote the total number of successes. Let us find $E\{S|X\}$ and $E\{X|S\}$.

We first compute $E\{S|X = n\}$. If $X = n$, then S is binomial with parameters n and p, and $E\{S|X = n\} = pn$. Thus $E\{S|X\} = pX$.

To compute $E\{X|S\}$, we need to compute $E\{X|S = k\}$; to do this we first compute $P(X = n|S = k)$:

$$\begin{aligned} P(X = n|S = k) &= \frac{P(S = k|X = n)P(X = n)}{P(S = k)} \\ &= \frac{\binom{n}{k}p^k(1-p)^{n-k}\left(\frac{\lambda^n}{n!}\right)e^{-\lambda}}{\sum_{m \geq k}\binom{m}{k}p^k(1-p)^{m-k}\left(\frac{\lambda^m}{m!}\right)e^{-\lambda}} \\ &= \frac{((1-p)\lambda)^{n-k}}{(n-k)!}e^{-(1-p)\lambda} \end{aligned}$$

for $n \geq k$. Thus,

$$E\{X|S = k\} = \sum_{n \geq k} n \frac{((1-p)\lambda)^{n-k}}{(n-k)!}e^{-(1-p)\lambda} = k + (1-p)\lambda,$$

hence,

$$E\{X|S\} = S + (1-p)\lambda.$$

Finally, one can check directly that $E\{S\} = E\{E\{S|X\}\}$; also this follows from Theorem 23.3 below. Therefore, we also have that

$$E\{S\} = pE\{X\} = p\lambda.$$

Next we wish to consider the general case: that is, we wish to treat $E\{Y|X\}$ where X is no longer assumed to take only countably many values. The preceding approach does not work, because the events $\{X = x\}$ in general have probability zero. Nevertheless we found in the countable case that $E\{Y|X\} = f(X)$ for a function f, and it is this idea that extends to the general case, with the aid of the next theorem. Let us recall a definition already given in Chapter 10:

Definition 23.3. *Let* $X: (\Omega, \mathcal{A}) \to (\mathbf{R}^n, \mathcal{B}^n)$ *be measurable. The σ-algebra generated by X is* $\sigma(X) = X^{-1}(\mathcal{B}^n)$ *(it is a σ-algebra: see the proof of Theorem 8.1), which is also given by*

$$\sigma(X) = \left\{ A \subset \Omega : X^{-1}(B) = A, \text{ for some } B \in \mathcal{B}^n \right\}.$$

Theorem 23.2. *Let X be an $\mathbf{R}^n$ valued random variable and let Y be an $\mathbf{R}$-valued random variable. Y is measurable with respect to $\sigma(X)$ if and only if there exists a Borel measurable function f on $\mathbf{R}^n$ such that $Y = f(X)$.*

Proof. Suppose such a function f exists. Let $B \in \mathcal{B}$. Then $Y^{-1}(B) = X^{-1}(f^{-1}(B))$. But $\Lambda = f^{-1}(B) \in \mathcal{B}^n$, whence $X^{-1}(\Lambda) \in \sigma(X)$ (alternatively, see Theorem 8.2).

Next suppose $Y^{-1}(B) \in \sigma(X)$, for each $B \in \mathcal{B}$. Suppose first $Y = \sum_{i=1}^k a_i 1_{A_i}$ for some $k < \infty$, with the a_i's all distinct and the A_i's pairwise disjoint. Then $A_i \in \sigma(X)$, hence there exists $B_i \in \mathcal{B}^n$ such that $A_i = X^{-1}(B_i)$. Let $f(x) = \sum_{i=1}^k a_i 1_{B_i}(x)$, and we have $Y = f(X)$, with f Borel measurable: so the result is proved for every simple r.v. Y which is $\sigma(X)$-measurable. If Y is next assumed only positive, it can be written $Y = \lim_{n \to \infty} Y_n$, where Y_n are simple and non-decreasing in n. (See for example such a construction in Chapter 9.) Each Y_n is $\sigma(X)$ measurable and also $Y_n = f_n(X)$ as we have just seen. Set $f(x) = \limsup_{n \to \infty} f_n(x)$. Then

$$Y = \lim_{n \to \infty} Y_n = \lim_n f_n(X).$$

But

$$(\limsup_{n \to \infty} f_n)(X) = \limsup_n (f_n(X)).$$

and since $\limsup_{n \to \infty} f_n(x)$ is Borel measurable, we are done.

For general Y, we can write $Y = Y^+ - Y^-$, and we are reduced to the preceding case. $\qquad\square$

In what follows, let $(\Omega, \mathcal{A}, P)$ be a fixed and given probability space, and let $X : \Omega \to \mathbf{R}^n$. The space $\mathcal{L}^2(\Omega, \mathcal{A}, P)$ is the space of all random variables Y such that $E\{Y^2\} < \infty$. If we identify all random variables that are equal a.s., we get the space $L^2(\Omega, \mathcal{A}, P)$. We can define an inner product (or "scalar product") by

$$\langle Y, Z \rangle = E\{YZ\}.$$

Then $L^2(\Omega, \mathcal{A}, P)$ is a Hilbert space, as we saw in Chapter 22. Since $\sigma(X) \subset \mathcal{A}$, the set $L^2(\Omega, \sigma(X), P)$ is also a Hilbert space, and it is a (closed) Hilbert subspace of $L^2(\Omega, \mathcal{A}, P)$. (Note that $L^2(\Omega, \sigma(X), P)$ has the same inner product as does $L^2(\Omega, \mathcal{A}, P)$.)

Definition 23.4. *Let $Y \in L^2(\Omega, \mathcal{A}, P)$. Then* the conditional expectation of Y given X *is the unique element $\hat{Y}$ in $L^2(\Omega, \sigma(X), P)$ such that*

$$E\{\hat{Y}Z\} = E\{YZ\} \text{ for all } Z \in L^2(\Omega, \sigma(X), P). \tag{23.2}$$

We write

$$E\{Y|X\}$$

for the conditional expectation of Y given X, namely $\hat{Y}$.

Note that $\hat{Y}$ is simply the Hilbert space projection of Y on the closed linear subspace $L^2(\Omega, \sigma(X), P)$ of $L^2(\Omega, \mathcal{A}, P)$: this is a consequence of Corollary 22.1 (or Exercise 23.4), and thus the conditional expectation does exist.

Observe that since $E\{Y|X\}$ is $\sigma(X)$ measurable, by Theorem 23.2 there exists a Borel measurable f such that $E\{Y|X\} = f(X)$. Therefore (23.2) is equivalent to

$$E\{f(X)g(X)\} = E\{Yg(X)\} \tag{23.3}$$

for each Borel g such that $g(X) \in \mathcal{L}^2$.

Next let us replace $\sigma(X)$ with simply a σ-algebra $\mathcal{G}$ with $\mathcal{G} \in \mathcal{A}$. Then $L^2(\Omega, \mathcal{G}, P)$ is a sub-Hilbert space of $L^2(\Omega, \mathcal{A}, P)$, and we can make an analogous definition:

Definition 23.5. *Let $Y \in L^2(\Omega, \mathcal{A}, P)$ and let $\mathcal{G}$ be a sub σ-algebra of $\mathcal{A}$. Then* the conditional expectation of Y given $\mathcal{G}$ *is the unique element $E\{Y|\mathcal{G}\}$ of $L^2(\Omega, \mathcal{G}, P)$ such that*

$$E\{YZ\} = E\{E\{Y|\mathcal{G}\}Z\} \tag{23.4}$$

for all $Z \in L^2(\Omega, \mathcal{G}, P)$.

Important Note: The conditional expectation is an element of L^2, that is an "equivalence class" of random variables. Thus any statement like $E\{Y|\mathcal{G}\} \geq 0$ or $E\{Y|\mathcal{G}\} = Z$, etc... should be understood with an implicit "almost surely" qualifier, or equivalently as such: there is a "version" of $E\{Y|\mathcal{G}\}$ that is positive, or equal to Z, etc...

Theorem 23.3. *Let $Y \in L^2(\Omega, \mathcal{A}, P)$ and $\mathcal{G}$ be a sub σ-algebra of $\mathcal{A}$.*

a) *If $Y \geq 0$ then $E\{Y|\mathcal{G}\} \geq 0$;*
b) *If $\mathcal{G} = \sigma(X)$ for some random variable X, there exists a Borel measurable function f such that $E\{Y|\mathcal{G}\} = f(X)$;*
c) *$E\{E\{Y|\mathcal{G}\}\} = E\{Y\}$;*
d) *The map $Y \to E\{Y|\mathcal{G}\}$ is linear.*

Proof. Property (b) we proved immediately preceding the theorem. For (c) we need only to apply (23.4) with $Z = 1$. Property (d) follows from (23.4) as well: if U, V are in L^2, then

$$E\{(U + \alpha V)Z\} = E\{UZ\} + \alpha E\{VZ\}$$
$$= E\{E\{U|\mathcal{G}\}Z\} + \alpha E\{E\{V|\mathcal{G}\}Z\}$$
$$= E\{(E\{U|\mathcal{G}\} + \alpha E\{V|\mathcal{G}\})Z\},$$

and thus $E\{U + \alpha V|\mathcal{G}\} = E\{U|\mathcal{G}\} + \alpha E\{V|\mathcal{G}\}$ by uniqueness (alternatively, as said before, $E\{Y|\mathcal{G}\}$ is the projection of Y on the subspace $L^2(\Omega, \mathcal{G}, P)$, and projections have been shown to be linear in Corollary 22.2).

Finally for (a) we again use (23.4) and take Z to be $1_{\{E\{Y|\mathcal{G}\}<0\}}$, assuming $Y \geq 0$ a.s. Then $E\{YZ\} \geq 0$ since both Y and Z are nonnegative, but

$$E\{E\{Y|\mathcal{G}\}Z\} = E\{E\{Y|\mathcal{G}\}1_{\{E\{Y|\mathcal{G}\}<0\}}\} < 0 \quad \text{if} \quad P(\{E\{Y|\mathcal{G}\} < 0\}) > 0.$$

This violates (23.3), so we conclude $P(\{E\{Y|\mathcal{G}\} < 0\}) = 0$.   □

Remark 23.2. As one can see from Theorem 23.3, the key property of conditional expectation is the property *(23.4)*; our only use of Hilbert space projection was to show that the conditional expectation exists.

We now wish to extend the conditional expectation of Definition 23.4 to random variables in L^1, not just random variables in L^2. Here the technique of Hilbert space projection is no longer available to us.

Once again let $\mathcal{L}^1(\Omega, \mathcal{A}, P)$ be the space of all L^1 random variables; we identify all random variables that are equal a.s. and we get the (Banach) space $L^1(\Omega, \mathcal{A}, P)$. Analogously, let $L^+(\Omega, \mathcal{A}, P)$ be all nonnegative random variables, again identifying all a.s. equal random variables. We allow random variables to assume the value $+\infty$.

Lemma 23.1. *Let $Y \in L^+(\Omega, \mathcal{A}, P)$ and let $\mathcal{G}$ be a sub σ-algebra of $\mathcal{A}$. There exists a unique element $E\{Y|\mathcal{G}\}$ of $L^+(\Omega, \mathcal{G}, P)$ such that*

$$E\{YX\} = E\{E\{Y|\mathcal{G}\}X\} \tag{23.5}$$

for all X in $L^+(\Omega, \mathcal{G}, P)$ and this conditional expectation agrees with the one in Definition 23.5 if further $Y \in L^2(\Omega, \mathcal{A}, P)$. Moreover, if $0 \leq Y \leq Y'$, then

$$E\{Y|\mathcal{G}\} \leq E\{Y'|\mathcal{G}\}. \tag{23.6}$$

Proof. If Y is in $L^2(\Omega, \mathcal{A}, P)$ and positive, we define $E\{Y|\mathcal{G}\}$ as in Definition 23.5. If X in $L^+(\Omega, \mathcal{G}, P)$ then $X_n = X \wedge n$ is square-integrable. Hence the Monotone Convergence Theorem (applied twice) and (23.5) yield

$$E\{YX\} = \lim_n E\{YX_n\}$$
$$= \lim_n E\{E\{Y|\mathcal{G}\}X_n\}$$
$$= E\{E\{Y|\mathcal{G}\}X\} \tag{23.7}$$

and (23.5) holds for all positive X.

Let now Y be in $L^+(\Omega, \mathcal{A}, P)$. Each $Y_m = Y \wedge m$ is bounded and hence in L^2, and by Theorem 23.3, conditional expectation on L^2 is a positive operator, so $E\{Y \wedge m | \mathcal{G}\}$ is increasing; therefore the following limit exists and we can set

$$E\{Y | \mathcal{G}\} = \lim_{m \to \infty} E\{Y_m | \mathcal{G}\}. \tag{23.8}$$

If $X \in L^+(\Omega, \mathcal{G}, P)$, we apply the Monotone Convergence Theorem several times as well as (23.8) to deduce that:

$$E\{YX\} = \lim_m E\{Y_m X\}$$
$$= E\left\{\lim_m E\{Y_m | \mathcal{G}\} X\right\}$$
$$= E\{E\{Y | \mathcal{G}\} X\}.$$

Furthermore if $Y \leq Y'$ we have $Y \wedge m \leq Y' \wedge m$ for all m, hence $E\{Y \wedge m | \mathcal{G}\} \leq E\{Y' \wedge m | \mathcal{G}\}$ as well by Theorem 23.3(a). Therefore (23.6) holds.

It remains to establish the uniqueness of $E\{Y | \mathcal{G}\}$ as defined above. Let U and V be two versions of $E\{Y | \mathcal{G}\}$ and let $\Lambda_n = \{U < V \leq n\}$ and suppose $P(\Lambda_n) > 0$. Note that $\Lambda_n \in \mathcal{G}$. We then have

$$E\{Y 1_{\Lambda_n}\} = E\{U 1_{\Lambda_n}\} = E\{V 1_{\Lambda_n}\},$$

since $E\{Y 1_{\Lambda}\} = E\{E\{Y | \mathcal{G}\} 1_{\Lambda}\}$ for all $\Lambda \in \mathcal{G}$ by (23.7). Further, $0 \leq U 1_{\Lambda_n} \leq V 1_{\Lambda_n} \leq n$, and $P(\Lambda_n) > 0$ implies that the r.v. $V 1_{\Lambda_n}$ and $U 1_{\Lambda_n}$ are not a.s. equal: we deduce that $E\{U 1_{\Lambda}\} < E\{V 1_{\Lambda}\}$, whence a contradiction. Therefore $P(\Lambda_n) = 0$ for all n, and since $\{U > V\} = \cup_{n \geq 1} \Lambda_n$ we get $P\{U < V\}) = 0$; analogously $P(\{V > U\}) = 0$, and we have uniqueness. $\qquad \square$

Theorem 23.4. *Let $Y \in L^1(\Omega, \mathcal{A}, P)$ and let $\mathcal{G}$ be a sub σ-algebra of $\mathcal{A}$. There exists a unique element $E\{Y | \mathcal{G}\}$ of $L^1(\Omega, \mathcal{G}, P)$ such that*

$$E\{YX\} = E\{E\{Y | \mathcal{G}\} X\} \tag{23.9}$$

for all bounded $\mathcal{G}$-measurable X and this conditional expectation agrees with the one in Definition 23.5 (resp. Lemma 23.1) when further $Y \in L^2(\Omega, \mathcal{A}, P)$ (resp. $Y \geq 0$), and satisfies

a) *If $Y \geq 0$ then $E\{Y | \mathcal{G}\} \geq 0$;*
b) *The map $Y \to E\{Y | \mathcal{G}\}$ is linear.*

Proof. Since Y is in L^1, we can write

$$Y = Y^+ - Y^-$$

where $Y^+ = \max(Y, 0)$ and $Y^- = -\min(Y, 0)$: moreover Y^+ and Y^- are also in $L^1(\Omega, \mathcal{G}, P)$. Next set

$$E\{Y | \mathcal{G}\} = E\{Y^+ | \mathcal{G}\} - E\{Y^- | \mathcal{G}\}.$$

This formula makes sense: indeed the r.v. Y^+ and Y^-, hence $E\{Y^+|\mathcal{G}\}$ and $E\{Y^-|\mathcal{G}\}$ as well by Theorem 23.3(c), are integrable, hence a.s. finite. That $E\{Y|\mathcal{G}\}$ satisfies (23.9) follows from Lemma 23.1. For uniqueness, let U, V be two versions of $E\{Y|\mathcal{G}\}$, and let $\Lambda = \{U < V\}$. Then $\Lambda \in \mathcal{G}$, so 1_Λ is bounded and $\mathcal{G}$-measurable. Then $E\{Y1_\Lambda\} = E\{E\{Y|\mathcal{G}\}1_\Lambda\} = E\{U1_\Lambda\} = E\{V1_\Lambda\}$. But if $P(\Lambda) > 0$, then $E\{U1_\Lambda\} < E\{V1_\Lambda\}$, which is a contradiction. So $P(\Lambda) = 0$ and analogously $P(\{V < U\}) = 0$ as well.

The final statements are trivial consequences of the previous definition of $E\{Y|\mathcal{G}\}$ and of Lemma 23.1 and Theorem 23.3.   □

Example: Let (X, Z) be real-valued random variables having a joint density $f(x, z)$. Let g be a bounded function and let

$$Y = g(Z).$$

We wish to compute $E\{Y|X\} = E\{g(Z)|X\}$. Recall that X has density f_X given by

$$f_X(x) = \int f(x, z)dz$$

and we defined in Chapter 12 (see Theorem 12.2) a *conditional density for Z* given $X = x$ by:

$$f_{X=x}(z) = \frac{f(x, z)}{f_X(x)},$$

whenever $f_X(x) \neq 0$. Next consider

$$h(x) = \int g(z)f_{X=x}(z)dz.$$

We then have, for any bounded Borel function $k(x)$:

$$\begin{aligned}
E\{h(X)k(X)\} &= \int h(x)k(x)f_X(x)dx \\
&= \int\int g(z)f_{X=x}(z)dz\, k(x)f_X(x)dx \\
&= \int\int g(z)\frac{f(x, z)}{f_X(x)}k(x)f_X(x)dz\, dx \\
&= \int\int g(z)k(x)f(x, z)dz\, dx \\
&= E\{g(Z)k(X)\} = E\{Yk(X)\}.
\end{aligned}$$

Therefore by (23.9) we have that

$$E\{Y|X\} = h(X).$$

This gives us an explicit way to calculate conditional expectations in the case when we have densities.

Theorem 23.5. *Let Y be a positive or integrable r.v. on $(\Omega, \mathcal{F}, P)$. Let $\mathcal{G}$ be a sub σ-algebra. Then $E\{Y|\mathcal{G}\} = Y$ if and only if Y is $\mathcal{G}$-measurable.*

Proof. This is trivial from the definition of conditional expectation.   □

Theorem 23.6. *Let $Y \in L^1(\Omega, \mathcal{A}, P)$ and suppose X and Y are independent. Then*
$$E\{Y|X\} = E\{Y\}.$$

Proof. Let g be bounded Borel. Then $E\{Yg(X)\} = E\{Y\}E\{g(X)\}$ by independence. Thus taking $f(x) = E\{Y\}$ for all x (the constant function) in Theorem 23.2, we have the result by (23.9).   □

Theorem 23.7. *Let X, Y be random variables on $(\Omega, \mathcal{A}, P)$, let $\mathcal{G}$ be a sub σ-algebra of $\mathcal{A}$, and suppose that X is $\mathcal{G}$-measurable. In the two following cases:*

a) *the variables X, Y and XY are integrable,*
b) *the variables X and Y are positive,*

we have
$$E\{XY|\mathcal{G}\} = XE\{Y|\mathcal{G}\}.$$

Proof. Assume first (b). For any $\mathcal{G}$-measurable positive r.v. Z we have
$$E\{XYZ\} = E\{XZE\{Y|\mathcal{G}\}\}$$

by (23.5). Since $XE\{Y|\mathcal{G}\}$ is also $\mathcal{G}$-measurable, we deduce the result by another application of the characterization (23.5).

In case (a), we observe that X^+Y^+, X^-Y^+, X^+Y^- and X^-Y^- are all integrable and positive. Then $E\{X^+Y^+|\mathcal{G}\} = X^+E\{Y^+|\mathcal{G}\}$ by what precedes, and similarly for the other three products, and all these quantities are finite. It remains to apply the linearity of the conditional expectation and the property $XY = X^+Y^+ + X^-Y^- - X^+Y^- - X^-Y^+$.   □

Let us note the important observation that the principal convergence theorems also hold for conditional expectations (we choose to emphasize below the fact that all statements about conditional expectations are "almost sure"):

Theorem 23.8. *Let $(Y_n)_{n \geq 1}$ be a sequence of r.v.'s on $(\Omega, \mathcal{A}, P)$ and let $\mathcal{G}$ be a sub σ-algebra of $\mathcal{A}$.*

a) *(Monotone Convergence.) If $Y_n \geq 0$, $n \geq 1$, and Y_n increases to Y a.s., then*
$$\lim_{n \to \infty} E\{Y_n|\mathcal{G}\} = E\{Y|\mathcal{G}\} \quad a.s.;$$

b) *(Fatou's Lemma.)* If $Y_n \geq 0$, $n \geq 1$, then

$$E\{\liminf_{n\to\infty} Y_n | \mathcal{G}\} \leq \liminf_{n\to\infty} E\{Y_n | \mathcal{G}\} \quad a.s.;$$

c) *(Lebesgue's dominated convergence theorem.)* If $\lim_{n\to\infty} Y_n = Y$ a.s. and $|Y_n| \leq Z$ $(n \geq 1)$ for some $Z \in L^1(\Omega, \mathcal{A}, P)$, then

$$\lim_{n\to\infty} E\{Y_n | \mathcal{G}\} = E\{Y | \mathcal{G}\} \quad a.s..$$

Proof. a) By (23.6) we have $E\{Y_{n+1} | \mathcal{G}\} \geq E\{Y_n | \mathcal{G}\}$ a.s., each n; hence $U = \lim_{n\to\infty} E\{Y_n | \mathcal{G}\}$ exists a.s. Then for all positive and $\mathcal{G}$-measurable r.v. X we have:

$$E\{UX\} = \lim_{n\to\infty} E\{E\{Y_n|\mathcal{G}\}X\}$$
$$= \lim_{n\to\infty} E\{Y_n X\}$$

by (23.5); and

$$= \lim_{n\to\infty} E\{YX\}$$

by the usual monotone convergence theorem. Thus $U = E\{Y|\mathcal{G}\}$, again by (23.5).

The proofs of (b) and (c) are analogous in a similar vein to the proofs of Fatou's lemma and the Dominated Convergence Theorem without conditioning. $\square$

We end with three useful inequalities.

Theorem 23.9 (Jensen's Inequality). *Let* $\varphi \colon \mathbf{R} \to \mathbf{R}$ *be convex, and let* X *and* $\varphi(X)$ *be integrable random variables. For any sub-σ-algebra* $\mathcal{G}$,

$$\varphi \circ E\{X|\mathcal{G}\} \leq E\{\varphi(X)|\mathcal{G}\}.$$

Proof. A result in real analysis is that if $\varphi : \mathbf{R} \to \mathbf{R}$ is convex, then $\varphi(x) = \sup_n(a_n x + b_n)$ for a countable collection of real numbers (a_n, b_n). Then

$$E\{a_n X + b_n | \mathcal{G}\} = a_n E\{X|\mathcal{G}\} + b_n.$$

But $E\{a_n X + b_n | \mathcal{G}\} \leq E\{\varphi(X)|\mathcal{G}\}$, hence $a_n E\{X|\mathcal{G}\} + b_n \leq E\{\varphi(X)|\mathcal{G}\}$, all n. Taking the supremum in n, we get the result. $\square$

Note that $\varphi(x) = x^2$ is of course convex, and thus as a consequence of Jensen's inequality we have

$$(E\{X|\mathcal{G}\})^2 \leq E\{X^2|\mathcal{G}\}.$$

An important consequence of Jensen's inequality is Hölder's inequality for random variables.

Theorem 23.10 (Hölder's Inequality). *Let X, Y be random variables with $E\{|X|^p\} < \infty$, $E\{|Y|^q\} < \infty$, where $p > 1$, and $\frac{1}{p} + \frac{1}{q} = 1$. Then*

$$|E\{XY\}| \le E\{|XY|\} \le E\{|X|^p\}^{\frac{1}{p}} E\{|Y|^q\}^{\frac{1}{q}}.$$

(Hence if $X \in L^p$ and $Y \in L^q$ with p, q as above, then the product XY belongs to L^1).

Proof. Without loss of generality we can assume $X \ge 0$, $Y \ge 0$ and $E\{X^p\} > 0$, since $E\{X^p\} = 0$ implies $X^p = 0$ a.s., thus $X = 0$ a.s. and there is nothing to prove. Let $C = E\{X^p\} < \infty$. Define a new probability measure Q by

$$Q(\Lambda) = \frac{1}{C} E\{1_\Lambda X^p\}$$

(compare with Exercises 9.5 and 9.7). Next define $Z = \frac{Y}{X^{p-1}} 1_{\{X>0\}}$. Since $\varphi(x) = |x|^p$ is convex, Jensen's inequality (Theorem 23.9) yields

$$(E_Q\{Z\})^q \le E\{Z^q\}.$$

Thus,

$$
\begin{aligned}
\frac{1}{C^q} E\{XY\}^q &= \frac{1}{C^q} E\left\{\frac{Y}{X^{p-1}} X^p\right\}^q \\
&= \left(E_Q\left\{\frac{Y}{X^{p-1}}\right\}\right)^q \\
&\le E_Q\left\{\left(\frac{Y}{X^{p-1}}\right)^q\right\} \\
&= \frac{1}{C} E\left\{\left(\frac{Y}{X^{p-1}}\right)^q X^p\right\} \\
&= \frac{1}{C} E\left\{Y^q \frac{1}{X^{(p-1)q}} X^p\right\},
\end{aligned}
$$

and $q = \frac{p}{p-1}$ while $(p-1)q = p$, hence

$$
\begin{aligned}
&= \frac{1}{C} E\left\{Y^q \frac{1}{X^p} X^p\right\} \\
&= \frac{1}{C} E\{Y^q\}.
\end{aligned}
$$

Thus

$$E\{XY\}^q \le C^{q-1} E\{Y^q\},$$

and taking q^{th} roots yields

$$E\{XY\} \le C^{\frac{q-1}{q}} E\{Y^q\}^{\frac{1}{q}}.$$

Since $\frac{q-1}{q} = \frac{1}{p}$ and $C = E\{X^p\}$, we have the result. $\qquad\square$

Corollary 23.1 (Minkowski's Inequality). *Let X, Y be random variables and $1 \leq p < \infty$ with $E\{|X|^p\} < \infty$ and $E\{|Y|^p\} < \infty$. Then*

$$E\{|X + Y|^p\}^{\frac{1}{p}} \leq E\{X^p\}^{\frac{1}{p}} + E\{Y^p\}^{\frac{1}{p}}.$$

Proof. If $p = 1$ the result is trivial. We therefore asume that $p > 1$. We use Hölder's inequality (Theorem 23.10). We have

$$E\{|X + Y|^p\} = E\left\{|X| |X + Y|^{p-1}\right\} + E\left\{|Y| |X + Y|^{p-1}\right\}$$
$$\leq E\{|X|^p\}^{\frac{1}{p}} E\{|X + Y|^{(p-1)q}\}^{\frac{1}{q}} + E\{|Y|^p\}^{\frac{1}{p}} E\{|X + Y|^{(p-1)q}\}^{\frac{1}{q}}.$$

But $(p-1)q = p$, and $\frac{1}{q} = 1 - \frac{1}{p}$, hence

$$= \left(E\{|X|^p\}^{\frac{1}{p}} + E\{|Y|^p\}^{\frac{1}{p}} \right) E\{|X + Y|^p\}^{1 - \frac{1}{p}}$$

and we have the result.   □

Minkowski's inequality allows one to define a norm (satisfying a triangle inequality) on the space L^p of equivalence classes (for the relation "equality a.s.") of random variables with $E\{|X|^p\} < \infty$.

Definition 23.6. *For X in L^p, define a norm by*

$$\|X\|_p = E\{|X^p|\}^{\frac{1}{p}}.$$

Note that Minkowski's inequality shows that L^p is a bonafide normed linear space. In fact it is even a *complete* normed linear space (called a "Banach space"). But for $p \neq 2$ it is not a Hilbert space: the norm is not associated with an inner product.

Exercises for Chapter 23

For Exercises 23.1–23.6, let Y be a positive or integrable random variable on the space $(\Omega, \mathcal{A}, P)$ and $\mathcal{G}$ be a sub σ-algebra of $\mathcal{A}$.

23.1 Show $|E\{Y|\mathcal{G}\}| \le E\{|Y||\mathcal{G}\}$.

23.2 Suppose $\mathcal{H} \subset \mathcal{G}$ where $\mathcal{H}$ is a sub σ-algebra of $\mathcal{G}$. Show that

$$E\{E\{Y|\mathcal{G}\}|\mathcal{H}\} = E\{Y|\mathcal{H}\}.$$

23.3 Show that $E\{Y|Y\} = Y$ a.s.

23.4 Show that if $|Y| \le c$ a.s. then $|E\{Y|\mathcal{G}\}| \le c$ a.s. also.

23.5 If $Y = \alpha$ a.s., with α a constant, show that $E\{Y|\mathcal{G}\} = \alpha$ a.s.

23.6 If Y is positive, show that $\{E\{Y|\mathcal{G}\} = 0\} \subset \{Y = 0\}$ and $\{Y = +\infty\} \subset \{E\{Y|\mathcal{G}\} = +\infty\}$ almost surely.

23.7 * Let X, Y be independent and let f be Borel such that $f(X,Y) \in L^1(\Omega, \mathcal{A}, P)$. Let

$$g(x) = \begin{cases} E\{f(x,Y)\} & \text{if } |E\{f(x,Y)\}| < \infty, \\ 0 & \text{otherwise.} \end{cases}$$

Show that g is Borel on $\mathbf{R}$ and that

$$E\{f(X,Y)|X\} = g(X).$$

23.8 Let Y be in $L^2(\Omega, \mathcal{A}, P)$ and suppose $E\{Y^2 \mid X\} = X^2$ and $E\{Y \mid X\} = X$. Show $Y = X$ a.s.

23.9 * Let Y be an exponential r.v. such that $P(\{Y > t\}) = e^{-t}$ for $t > 0$. Calculate $E\{Y \mid Y \wedge t\}$, where $Y \wedge t = \min(t, Y)$.

23.10 (Chebyshev's inequality). Prove that for $X \in L^2$ and $a > 0$, $P(|X| \ge a|\mathcal{G}) \le \frac{E\{X^2|\mathcal{G}\}}{a^2}$ (by $P(A|\mathcal{G})$ one means the conditional expectation $E(1_A|\mathcal{G})$).

23.11 (Cauchy-Schwarz). For X, Y in L^2 show

$$(E\{XY|\mathcal{G}\})^2 \le E\{X^2|\mathcal{G}\}E\{Y^2|\mathcal{G}\}.$$

23.12 Let $X \in L^2$. Show that

$$E\{(X - E\{X|\mathcal{G}\})^2\} \le E\{(X - E\{X\})^2\}.$$

23.13 Let $p \ge 1$ and $r \ge p$. Show that $L^p \supset L^r$, for expectation with respect to a probability measure.

23.14 * Let Z be defined on $(\Omega, \mathcal{F}, P)$ with $Z \geq 0$ and $E\{Z\} = 1$. Define a new probability Q by $Q(\Lambda) = E\{1_\Lambda Z\}$. Let $\mathcal{G}$ be a sub σ-algebra of $\mathcal{F}$, and let $U = E\{Z|\mathcal{G}\}$. Show that $E_Q\{X|\mathcal{G}\} = \frac{E\{XZ|\mathcal{G}\}}{U}$, for any bounded $\mathcal{F}$-measurable random variable X. (Here $E_Q\{X|\mathcal{G}\}$ denotes the conditional expectation of X relative to the probability measure Q.)

23.15 Show that the normed linear space L^p is complete for each p, $1 \leq p < \infty$. (*Hint:* See the proof of Theorem 22.2.)

23.16 Let $X \in L^1(\Omega, \mathcal{F}, P)$ and let $\mathcal{G}, \mathcal{H}$ be sub σ-algebras of $\mathcal{F}$. Moreover let $\mathcal{H}$ be independent of $\sigma(\sigma(X), \mathcal{G})$. Show that $E\{X|\sigma(\mathcal{G}, \mathcal{H})\} = E\{X|\mathcal{G}\}$.

23.17 Let $(X_n)_{n \geq 1}$ be independent and identically distributed and in L^1 and let $S_n = \sum_{i=1}^n X_i$ and $\mathcal{G}_n = \sigma(S_n, S_{n+1}, \ldots)$. Show that $E\{X_1|\mathcal{G}_n\} = E\{X_1 \mid S_n\}$ and also $E\{X_j|\mathcal{G}_n\} = E\{X_j \mid S_n\}$ for $1 \leq j \leq n$. Also show that $E\{X_j|\mathcal{G}_n\} = E\{X_1|S_n\}$ for $1 \leq j \leq n$ (*Hint:* Use Exercise 23.16.)

23.18 Let $X_1, X_2, \ldots, X_n$ be independent and identically distributed and in L_1. Show that for each j, $1 \leq j \leq n$, we have $E\{X_j| \sum_{i=1}^n X_i\} = \frac{1}{n} \sum_{i=1}^n X_i$. (*Hint:* Use Theorem 23.2 and the symmetry coming from the i.i.d. hypothesis.)

24 Martingales

We begin by recalling the Strong Law of Large Numbers (Theorem 20.1): if $(X_n)_{n\geq 1}$ are i.i.d. with $E\{X_n\} = \mu$ and $\sigma^2_{X_n} < \infty$, and if $S_n = \sum_{j=1}^{n} X_j$, then $\lim_{n\to\infty} \frac{S_n}{n} = \mu$ a.s. Note that since the X_n are all independent, the limit must be constant a.s. as a consequence of the tail event zero–one law (Theorem 10.6). It is interesting to study sequences converging to limits that are random variables, not just constant.

Let us rewrite the sequence as

$$\lim_{n\to\infty} \frac{S_n - n\mu}{n} = 0 \quad \text{a.s.}$$

A key property of this sequence is that if $\mathcal{F}_n = \sigma\{S_k; k \leq n\}$, then

$$E\{S_{n+1} - (n+1)\mu | \mathcal{F}_n\} = S_n - n\mu, \tag{24.1}$$

as will be seen in Example 24.1 below. It is property (24.1) that is the key to the study of more general types of convergence, where we relax the independence assumption.

We assume given and fixed both a probability space $(\Omega, \mathcal{F}, P)$ and an increasing sequence of σ-algebras $(\mathcal{F}_n)_{n\geq 0}$, having the property that $\mathcal{F}_n \subset \mathcal{F}_{n+1} \subset \mathcal{F}$, all $n \geq 0$.

Definition 24.1. *A sequence of random variables $(X_n)_{n\geq 0}$ is called a martingale, or an $(\mathcal{F}_n)$ martingale, if*

(i) $E\{|X_n|\} < \infty$, *each n;*
(ii) X_n *is $\mathcal{F}_n$ measurable, each n;*
(iii) $E\{X_n | \mathcal{F}_m\} = X_m$ *a.s., each $m \leq n$.*

Note that (ii) is "almost" implied by (iii), which yields that X_m is a.s. equal to an $\mathcal{F}_m$ measurable random variable.

Example 24.1. Let $(X_n)_{n\geq 1}$ be independent with $E\{|X_n|\} < \infty$ and $E\{X_n\} = 0$, all n. For $n \geq 1$ let $\mathcal{F}_n = \sigma\{X_k; k \leq n\}$ and $S_n = \sum_{k=1}^{n} X_k$. For $n = 0$ let $\mathcal{F}_0 = \{\phi, \Omega\}$ be the "trivial" σ-algebra and $S_0 = 0$. Then $(S_n)_{n\geq 0}$ is an $(\mathcal{F}_n)_{n\geq 0}$ martingale, since

$$E\{S_n|\mathcal{F}_m\} = E\{S_m + (S_n - S_m)|\mathcal{F}_m\}$$
$$= S_m + E\{S_n - S_m|\mathcal{F}_m\}$$
$$= S_m + E\left\{\sum_{k=m+1}^{n} X_k|\mathcal{F}_m\right\}$$
$$= S_m + \sum_{k=m+1}^{n} E\{X_k\}$$
$$= S_m.$$

When the variables X_n have $\mu = E\{X_n\} \neq 0$, then using $X_n - \mu$ instead of X_n above we obtain similarly that $(S_n - n\mu)_{n\geq0}$ is an $(\mathcal{F}_n)_{n\geq0}$ martingale.

Example 24.2. Let Y be $\mathcal{F}$–measurable with $E\{|Y|\} < \infty$ and define

$$X_n = E\{Y|\mathcal{F}_n\}.$$

Then $E\{|X_n|\} \leq E\{|Y|\} < \infty$ and for $m \leq n$,

$$E\{X_n|\mathcal{F}_m\} = E\{E\{Y|\mathcal{F}_n\}|\mathcal{F}_m\}$$
$$= E\{Y|\mathcal{F}_m\}$$
$$= X_m$$

(see Exercises 23.1 and 23.2).

Definition 24.2. *A martingale $X = (X_n)_{n\geq0}$ is said to be* closed *by a random variable Y if $E\{|Y|\} < \infty$ and $X_n = E\{Y|\mathcal{F}_n\}$, each n.*

Example 24.2 shows that any r.v. $Y \in \mathcal{F}$ with $E\{|Y|\} < \infty$ gives an example of a closed martingale by taking $X_n = E\{Y|\mathcal{F}_n\}$, $n \geq 0$.

An important property of martingales is that a martingale has constant expectation:

Theorem 24.1. *If $(X_n)_{n\geq0}$ is a martingale, then $n \to E\{X_n\}$ is constant. That is, $E\{X_n\} = E\{X_0\}$, all $n \geq 0$.*

Proof. $E\{X_n\} = E\{E\{X_n|\mathcal{F}_0\}\} = E\{X_0\}$. □

The converse of Theorem 24.1 is not true, but there is a partial converse using stopping times (see Theorem 24.7).

Definition 24.3. *A random variable $T: \Omega \to \overline{\mathbf{N}} = \mathbf{N} \cup \{+\infty\}$ is called a* stopping time *if $\{T \leq n\} \in \mathcal{F}_n$, for all n.*

Any constant r.v. equal to an integer, or to $+\infty$, is a stopping time. Stopping times are often more useful than fixed times. They can be thought of as the time when a given random event happens, with the convention that it takes the value $+\infty$ if this event never happens. For example suppose

$(X_n)_{n\geq0}$ is a martingale and we are interested in the first time it is at least 12. Such a time will be random and can be expressed as

$$T = \begin{cases} \inf_{n\geq0}\{n : X_n \geq 12\} & \text{if } X_n \geq 12 \text{ for some } n \in \mathbf{N} \\ +\infty & \text{otherwise.} \end{cases}$$

That is,

$$T(\omega) = \inf_{n\geq0}\{n : X_n(\omega) \geq 12\}$$

if $X_n(\omega) \geq 12$ for some integer n, and $T(\omega) = +\infty$ if not. Note that the event $\{\omega : T(\omega) \leq n\}$ can be expressed as:

$$\{T \leq n\} = \bigcup_{k=0}^{n}\{X_k \geq 12\} \in \mathcal{F}_n$$

because $\{X_k \geq 12\} \in \mathcal{F}_k \subset \mathcal{F}_n$ if $k \leq n$. The term "stopping time" comes from gambling: a gambler can decide to stop playing at a random time (depending for example on previous gains or losses), but when he or she actually decides to stop, his or her decision is based upon the knowledge of what happened before and at that time, and obviously not on future outcomes: the reader can check that this corresponds to Definition 24.3.

Theorem 24.1 extends to bounded stopping times (a stopping time is T is *bounded* if there exists a constant c such that $P\{T \leq c\} = 1$). If T is a finite stopping time, we denote by X_T the r.v. $X_T(\omega) = X_{T(\omega)}(\omega)$; that is, it takes the value X_n whenever $T = n$.

Theorem 24.2. *Let T be a stopping time bounded by c and let $(X_n)_{n\geq0}$ be a martingale. Then $E\{X_T\} = E\{X_0\}$.*

Proof. We have $X_T(\omega) = \sum_{n=0}^{\infty} X_n(\omega)1_{\{T(\omega)=n\}}$. Therefore, assuming without loss of generality that c is itself an integer,

$$E\{X_T\} = E\left\{\sum_{n=0}^{\infty} X_n 1_{\{T=n\}}\right\}$$

$$= E\left\{\sum_{n=0}^{c} X_n 1_{\{T=n\}}\right\}$$

$$= \sum_{n=0}^{c} E\{X_n 1_{\{T=n\}}\}.$$

Since $\{T = n\} = \{T \leq n\} \setminus \{T \leq n-1\}$ we see $\{T = n\} \in \mathcal{F}_n$, and we obtain

$$= \sum_{n=0}^{c} E\{E\{X_c|\mathcal{F}_n\}1_{\{T=n\}}\}$$

$$= \sum_{n=0}^{c} E\{X_c 1_{\{T=n\}}\}$$

$$= E\left\{X_c \sum_{n=0}^{c} 1_{\{T=n\}}\right\}$$

$$= E\{X_c\} = E\{X_0\},$$

with the last equality by Theorem 24.1.   □

The σ-algebra $\mathcal{F}_n$ can be thought of as representing observable events up to and including time n. We wish to create an analogous notion of observable events up to a stopping time T.

Definition 24.4. *Let T be a stopping time. The stopping time σ-algebra $\mathcal{F}_T$ is defined to be*

$$\mathcal{F}_T = \{A \in \mathcal{F} : A \cap \{T \leq n\} \in \mathcal{F}_n, \text{ all } n\}.$$

For the above definition to make sense, we need a minor result:

Theorem 24.3. *For T a stopping time, $\mathcal{F}_T$ is a σ-algebra.*

Proof. Clearly ϕ and Ω are in $\mathcal{F}_T$. If $A \in \mathcal{F}_T$, then

$$A^c \cap \{T \leq n\} = \{T \leq n\} \setminus (A \cap \{T \leq n\}),$$

and thus $A^c \in \mathcal{F}_T$. Also if $(A_i)_{i \geq 1}$ are in $\mathcal{F}_T$, then

$$\left(\bigcup_{i=1}^{\infty} A_i\right) \cap \{T \leq n\} = \bigcup_{i=1}^{\infty} (A_i \cap \{T \leq n\}) \in \mathcal{F}_n,$$

hence $\mathcal{F}_T$ is closed under complements and countable unions; thus it is a σ-algebra.   □

Theorem 24.4. *Let S, T be stopping times, with $S \leq T$. Then $\mathcal{F}_S \subset \mathcal{F}_T$.*

Proof. Since $S \leq T$ we have $\{T \leq n\} \subset \{S \leq n\}$. Therefore if $A \in \mathcal{F}_S$, then:

$$A \cap \{T \leq n\} = A \cap \{S \leq n\} \cap \{T \leq n\};$$

but $A \cap \{S \leq n\} \in \mathcal{F}_n$ and $\{T \leq n\} \in \mathcal{F}_n$, so $A \cap \{T \leq n\} \in \mathcal{F}_n$, hence $A \in \mathcal{F}_T$.   □

Next assume that $(X_n)_{n \geq 0}$ is a sequence of random variables with X_n being $\mathcal{F}_n$ measurable, each n. Let T be a stopping time with $P(T < \infty) = 1$. Then $X_T = \sum_{n=0}^{\infty} X_n 1_{\{T=n\}}$, and we have:

Theorem 24.5. *X_T is $\mathcal{F}_T$-measurable.*

Proof. Let Λ be Borel and we want to show $\{X_T \in \Lambda\} \in \mathcal{F}_T$; that is, we need $\{X_T \in \Lambda\} \cap \{T \le n\} \in \mathcal{F}_n$. But

$$\{X_T \in \Lambda\} \cap \{T \le n\}$$
$$= \bigcup_{k=0}^{n} \{X_T \in \Lambda\} \cap \{T = k\}$$
$$= \bigcup_{k=0}^{n} \{X_k \in \Lambda\} \cap \{T = k\},$$

and $\{X_k \in \Lambda\} \cap \{T = k\} \in \mathcal{F}_k \subset \mathcal{F}_n$ for $k \le n$. $\square$

The next two theorems show that the martingale property holds at stopping times as well as fixed times. This is a surprisingly powerful result.

Theorem 24.6 (Doob's Optional Sampling Theorem). *Let* $X = (X_n)_{n \ge 0}$ *be a martingale and let* S, T *be stopping times bounded by a constant* c*, with* $S \le T$ *a.s. Then*

$$E\{X_T | \mathcal{F}_S\} = X_S \ \text{a.s.}$$

Proof. First $|X_T| \le \sum_{n=0}^{c} |X_n|$ is integrable (without loss of generality we can assume again that c is an integer), as well as X_S, and further X_S is $\mathcal{F}_S$-measurable by the previous theorem. So it remains to prove that $E\{X_T Z\} = E\{X_S Z\}$ for every bounded $\mathcal{F}_S$-measurable r.v. Z. By a standard argument it is even enough to prove that if $A \in \mathcal{F}_S$ then

$$E\{X_T 1_A\} = E\{X_S 1_A\}$$

(if this holds, then $E\{X_T Z\} = E\{X_S Z\}$ holds for simple Z by linearity, then for all $\mathcal{F}_S$-measurable and bounded Z by Lebesgue's Dominated Convergence Theorem).

So let $A \in \mathcal{F}_S$. Define a new random time R by

$$R(\omega) = S(\omega) 1_A(\omega) + T(\omega) 1_{A^c}(\omega).$$

Then R is a stopping time also: indeed,

$$\{R \le n\} = A \cap \{S \le n\}) \cup (A^c \cap \{T \le n\}),$$

and $A \cap \{S \le n\} \in \mathcal{F}_n$ because $A \in \mathcal{F}_S$. Since $A \in \mathcal{F}_S$ we have $A^c \in \mathcal{F}_S$ and so $A^c \in \mathcal{F}_T$ by Theorem 24.4. Thus $A^c \cap \{T \le n\} \in \mathcal{F}_n$ and we conclude $\{R \le n\} \in \mathcal{F}_n$ and R is a stopping time. Therefore $E\{X_R\} = E\{X_T\} = E\{X_0\}$ by Theorem 24.2. But

$$E\{X_R\} = E\{X_S 1_A + X_T 1_{A^c}\},$$
$$E\{X_T\} = E\{X_T 1_A + X_T 1_{A^c}\}$$

and subtracting yields

$$E\{X_S 1_A\} - E\{X_T 1_A\} = 0.$$

$\square$

We can now establish a partial converse of Theorem 24.1.

Theorem 24.7. *Let* $(X_n)_{n\geq 0}$ *be a sequence of random variables with* X_n *being* $\mathcal{F}_n$ *measurable, each* n. *Suppose* $E\{|X_n|\} < \infty$ *for each* n, *and* $E\{X_T\} = E\{X_0\}$ *for all bounded stopping times* T. *Then* X *is a martingale.*

Proof. Let $0 \leq m < n < \infty$, and let $\Lambda \in \mathcal{F}_m$. Define a random time T by:

$$T(\omega) = \begin{cases} m \text{ if } \omega \in \Lambda^c, \\ n \text{ if } \omega \in \Lambda \end{cases}$$

Then T is a stopping time, so

$$E\{X_0\} = E\{X_T\} = E\{X_m 1_{\Lambda^c} + X_n 1_\Lambda\}.$$

However also $E\{X_0\} = E\{X_m 1_{\Lambda^c} + X_m 1_\Lambda\}$. Subtraction yields $E\{X_n 1_\Lambda\} = E\{X_m 1_\Lambda\}$, or equivalently $E\{X_n | \mathcal{F}_m\} = X_m$ a.s. $\square$

Exercises for Chapter 24

In Problems 24.1–24.11 let S and T be stopping times for a sequence of σ–algebras $(\mathcal{F}_n)_{n \geq 0}$, with $\mathcal{F}_m \subset \mathcal{F}_n$ for $m \leq n$.

24.1 If $T \equiv n$, show that $\mathcal{F}_T = \mathcal{F}_n$.

24.2 Show that $S \wedge T = \min(S, T)$ is a stopping time.

24.3 Show that $S \vee T = \max(S, T)$ is a stopping time.

24.4 Show that $S + T$ is a stopping time.

24.5 Show that αT is a stopping time for $\alpha \geq 1$, α integer.

24.6 Show that $\mathcal{F}_{S \wedge T} \subset \mathcal{F}_T \subset \mathcal{F}_{S \vee T}$.

24.7 Show that T is a stopping time if and only if $\{T = n\} \in \mathcal{F}_n$, each $n \geq 0$.

24.8 Let $\Lambda \in \mathcal{F}_T$ and define

$$T_\Lambda(\omega) = \begin{cases} T(\omega) & \text{if } \omega \in \Lambda, \\ \infty & \text{if } \omega \notin \Lambda. \end{cases}$$

Show that T_Λ is another stopping time.

24.9 Show that T is $\mathcal{F}_T$–measurable.

24.10 Show that $\{S < T\}$, $\{S \leq T\}$, and $\{S = T\}$ are all in $\mathcal{F}_S \cap \mathcal{F}_T$.

24.11 * Show that

$$E\{E\{Y|\mathcal{F}_T\}|\mathcal{F}_S\} = E\{E\{Y|\mathcal{F}_S\}|\mathcal{F}_T\} = E\{Y|\mathcal{F}_{S \wedge T}\}.$$

24.12 Let $M = (M_n)_{n \geq 0}$ be a martingale with $M_n \in L^2$, each n. Let S, T be bounded stopping times with $S \leq T$. Show that M_S, M_T, are both in L^2, and show that

$$E\{(M_T - M_S)^2|\mathcal{F}_S\} = E\{M_T^2 - M_S^2|\mathcal{F}_S\},$$

and that

$$E\{(M_T - M_S)^2\} = E\{M_T^2\} - E\{M_S^2\}.$$

24.13 Let φ be convex and let $M = (M_n)_{n \geq 0}$ be a martingale. Show that $n \to E\{\varphi(M_n)\}$ is a nondecreasing function. (*Hint:* Use Jensen's inequality [Theorem 23.9].)

24.14 Let X_n be a sequence of random variables with $E\{X_n\} < \infty$ and $E\{X_n \mid \mathcal{F}_{n-1}\} = 0$ for each $n \geq 1$. Suppose further that X_n is $\mathcal{F}_n$-measurable, for each $n \geq 0$. Let $S_n = \sum_{k=0}^n X_k$. Show that $(S_n)_{n \geq 0}$ is a martingale for $(\mathcal{F}_n)_{n \geq 0}$.

25 Supermartingales and Submartingales

In Chapter 24 we defined a martingale via an equality for certain conditional expectations. If we replace that equality with an inequality we obtain super-martingales and submartingales. Once again $(\Omega, \mathcal{F}, P)$ is a probability space that is assumed given and fixed, and $(\mathcal{F}_n)_{n\geq 1}$ is an increasing sequence of σ-algebras.

Definition 25.1. *A sequence of random variables* $(X_n)_{n\geq 0}$ *is called a sub-martingale (respectively a* submartingale*) if*

(i) $E\{|X_n|\} < \infty$, *each* n;
(ii) X_n *is* $\mathcal{F}_n$-*measurable, each* n;
(iii) $E\{X_n|\mathcal{F}_m\} \geq X_m$ *a.s. (resp.* $\leq X_m$ *a.s.) each* $m \leq n$.

The sequence $(X_n)_{n\geq 0}$ is a martingale if and only if it is a submartingale *and* a supermartingale.

Theorem 25.1. *If* $(M_n)_{n\geq 0}$ *is a martingale, and if* φ *is convex and* $\varphi(M_n)$ *is integrable for each* n, *then* $(\varphi(M_n))_{n\geq 0}$ *is a submartingale.*

Proof. Let $m \leq n$. Then $E\{M_n|\mathcal{F}_m\} = M_m$ a.s., so $\varphi(E\{M_n|\mathcal{F}_m\}) = \varphi(M_m)$ a.s., and since φ is convex by Jensen's inequality (Theorem 23.9) we have

$$E\{\varphi(M_n)|\mathcal{F}_m\} \geq \varphi(E\{M_n|\mathcal{F}_m\}) = \varphi(M_m).$$

$\square$

Corollary 25.1. *If* $(M_n)_{n\geq 0}$ *is a martingale then* $X_n = |M_n|$, $n \geq 0$, *is a submartingale.*

Proof. $\varphi(x) = |x|$ is a convex, so apply Theorem 25.1. $\square$

Theorem 25.2. *Let* T *be a stopping time bounded by* $C \in \mathbf{N}$ *and let* $(X_n)_{n\geq 0}$ *be a submartingale. Then* $E\{X_T\} \leq E\{X_C\}$.

Proof. The proof is analogous to the proof of Theorem 24.2, so we omit it. $\square$

The next theorem shows a connection between submartingales and martingales.

Theorem 25.3 (Doob Decomposition). *Let* $X = (X_n)_{n \geq 0}$ *be a submartingale. There exists a martingale* $M = (M_n)_{n \geq 0}$ *and a process* $A = (A_n)_{n \geq 0}$ *with* $A_{n+1} \geq A_n$ *a.s. and* A_{n+1} *being* $\mathcal{F}_n$-*measurable, each* $n \geq 0$, *such that*

$$X_n = X_0 + M_n + A_n, \quad \text{with} \quad M_0 = A_0 = 0.$$

Moreover such a decomposition is a.s. unique.

Proof. Define $A_0 = 0$ and

$$A_n = \sum_{k=1}^{n} E\{X_k - X_{k-1} | \mathcal{F}_{k-1}\} \quad \text{for } n \geq 1.$$

Since X is a submartingale we have $E\{X_k - X_{k-1} | \mathcal{F}_{k-1}\} \geq 0$ each k, hence $A_{k+1} \geq A_k$ a.s., and also A_{k+1} being $\mathcal{F}_k$-measurable. Note also that

$$E\{X_n \mid \mathcal{F}_{n-1}\} - X_{n-1} = E\{X_n - X_{n-1} \mid \mathcal{F}_{n-1}\} = A_n - A_{n-1},$$

and hence

$$E\{X_n \mid \mathcal{F}_{n-1}\} - A_n = X_{n-1} - A_{n-1};$$

but $A_n \in \mathcal{F}_{n-1}$, so

$$E\{X_n - A_n | \mathcal{F}_{n-1}\} = X_{n-1} - A_{n-1}. \tag{25.1}$$

Letting $M_n = X_n - X_0 - A_n$ we have from (25.1) that M is a martingale and we have the existence of the decomposition.

As for uniqueness, suppose

$$X_n = X_0 + M_n + A_n, \quad n \geq 0,$$
$$X_n = X_0 + L_n + C_n, \quad n \geq 0,$$

are two such decompositions. Subtracting one from the other gives

$$L_n - M_n = A_n - C_n. \tag{25.2}$$

Since A_n, C_n are $\mathcal{F}_{n-1}$ measurable, $L_n - M_n$ is $\mathcal{F}_{n-1}$ measurable as well; therefore

$$L_n - M_n = E\{L_n - M_n | \mathcal{F}_{n-1}\} = L_{n-1} - M_{n-1} = A_{n-1} - C_{n-1} \quad \text{a.s.}$$

Continuing inductively we see that $L_n - M_n = L_0 - M_0 = 0$ a.s. since $L_0 = M_0 = 0$. We conclude that $L_n = M_n$ a.s., whence $A_n = C_n$ a.s. and we have uniqueness. $\qquad \square$

Corollary 25.2. *Let* $X = (X_n)_{n \geq 0}$ *be a supermartingale. There exists a unique decomposition*

$$X_n = X_0 + M_n - A_n, \quad n \geq 0$$

with $M_0 = A_0 = 0$, $(M_n)_{n \geq 0}$ *a martingale, and* A_k *being* $\mathcal{F}_{k-1}$-*measurable with* $A_k \geq A_{k-1}$ *a.s.*

Proof. Let $Y_n = -X_n$. Then $(Y_n)_{n \geq 0}$ is a submartingale. Let the Doob decomposition be

$$Y_n = Y_0 + L_n + C_n,$$

and then $X_n = X_0 - L_n - C_n$; set $M_n = -L_n$ and $A_n = C_n$, $n \geq 0$.   □

Exercises for Chapter 25

25.1 Show that $X = (X_n)_{n \geq 0}$ is a submartingale if and only if $Y_n = -X_n$, $n \geq 0$, is a supermartingale.

25.2 Show that if $X = (X_n)_{n \geq 0}$ is both a submartingale and a supermartingale, then X is a martingale.

25.3 Let $X = (X_n)_{n \geq 0}$ be a submartingale with Doob decomposition $X_n = X_0 + M_n + A_n$. Show that $E\{A_n\} < \infty$, each $n < \infty$.

25.4 Let $M = (M_n)_{n \geq 0}$ be a martingale with $M_0 = 0$ and suppose $E\{M_n^2\} < \infty$, each n. Show that $X_n = M_n^2$, $n \geq 0$, is a submartingale, and let $X_n = L_n + A_n$ be its Doob decomposition. Show that $E\{M_n^2\} = E\{A_n\}$.

25.5 Let M and A be as in Exercise 25.4. Show that $A_n - A_{n-1} = E\{(M_n - M_{n-1})^2 | \mathcal{F}_{n-1}\}$.

25.6 Let $X = (X_n)_{n \geq 0}$ be a submartingale. Show that if φ is convex and nondecreasing on $\mathbf{R}$ and if $\varphi(X_n)$ is integrable for each n, then $Y_n = \varphi(X_n)$ is also a submartingale.

25.7 Let $X = (X_n)_{n \geq 0}$ be an increasing sequence of integrable r.v., each X_n being $\mathcal{F}_n$-measurable. Show that X is a submartingale.

26 Martingale Inequalities

One of the reasons martingales have become central to probability theory is that their structure gives rise to some powerful inequalities. Our presentation follows Bass [1].

Once again $(\Omega, \mathcal{F}, P)$ is a probability space that is assumed given and fixed, and $(\mathcal{F}_n)_{n \geq 0}$ is an increasing sequence of σ-algebras. Let $M = (M_n)_{n \geq 0}$ be a sequence of integrable r.v.'s, each M_n being $\mathcal{F}_n$-measurable, and let $M_n^* = \sup_{j \leq n} |M_j|$. Note that M_n^* is an increasing process and a submartingale (see Exercise 25.7), since

$$E\{M_n^*\} \leq E\left\{\sum_{j=1}^n |M_j|\right\} < \infty.$$

By Markov's Inequality (Corollary 5.1)

$$P(M_n^* \geq \alpha) = E\{1_{\{M_n^* \geq \alpha\}}\} \leq \frac{E\{M_n^*\}}{\alpha}.$$

In the martingale case we can replace M_n^* with only $|M_n|$ on the right side.

Theorem 26.1 (Doob's First Martingale Inequality). *Let* $M = (M_n)_{n \geq 0}$ *be a martingale or a positive submartingale. Then*

$$P(M_n^* \geq \alpha) \leq \frac{E\{|M_n|\}}{\alpha}.$$

Proof. Let $T = \min\{j : |M_j| \geq \alpha\}$ (recall our convention that the minimum of an empty subset of $\mathbf{N}$ is $+\infty$). Since $\varphi(x) = |x|$ is convex and increasing on $\mathbf{R}_+$, we have that $|M_n|$ is a submartingale (by Corollary 25.1 if M is a martingale, or by Exercise 25.6 if M is a positive submartingale). The set $\{T \leq n, |M_T| \geq \alpha\}$ and $\{M_n^* \geq \alpha\}$ are equal, hence

$$P(M_n^* \geq \alpha) = P(T \leq n, |M_T| \geq \alpha) \leq E\left\{\frac{|M_T|}{\alpha} 1_{\{T \leq n\}}\right\},$$

and since $M_T = M_{T \wedge n}$ on $\{T \leq n\}$,

$$P(M_n^* \geq \alpha) \leq \frac{1}{\alpha} E\{|M_{T \wedge n}| 1_{\{T \leq n\}}\} \leq \frac{E\{|M_{T \wedge n}|\}}{\alpha} \leq \frac{E\{|M_n|\}}{\alpha}$$

by Theorem 25.2. $\qquad\square$

Before we prove Doob's L^p Martingale Inequalities we need a lemma which is interesting in its own right.

Lemma 26.1. *Let $X \geq 0$ be a random variable, $p > 0$, and $E\{X^p\} < \infty$. Then*

$$E\{X^p\} = \int_0^\infty p\lambda^{p-1} P(X > \lambda)d\lambda.$$

Proof. We have

$$\int_0^\infty p\lambda^{p-1} P(X > \lambda)d\lambda = \int_0^\infty p\lambda^{p-1} E\{1_{(X>\lambda)}\}d\lambda,$$

and by Fubini's Theorem (see Exercise 10.15)

$$= E\left\{\int_0^\infty p\lambda^{p-1} 1_{(X>\lambda)}d\lambda\right\} = E\left\{\int_0^X p\lambda^{p-1}d\lambda\right\} = E\{X^p\}.$$

$\square$

Theorem 26.2 (Doob's L^p Martingale Inequalities). *Let $M = (M_n)_{n\geq 0}$ be a martingale or a positive submartingale. Let $1 < p < \infty$. There exists a constant c depending only on p such that*

$$E\{(M_n^*)^p\} \leq cE\{|M_n|^p\}.$$

Proof. We give the proof in the martingale case. Since $\varphi(x) = |x|$ is convex we have $|M_n|$ is a submartingale as in Theorem 26.1. Let $X_n = M_n 1_{(|M_n|>\frac{\alpha}{2})}$. For n fixed define

$$Z_j = E\{X_n | \mathcal{F}_j\}, \quad 0 \leq j \leq n.$$

Note that Z_j, $0 \leq j \leq n$ is a martingale. Note further that $M_n^* \leq Z_n^* + \frac{\alpha}{2}$, since

$$\begin{aligned}
|M_j| &= |E\{M_n \mid \mathcal{F}_j\}| \\
&= |E\{M_n 1_{(|M_n|>\frac{\alpha}{2})} + M_n 1_{(|M_n|\leq\frac{\alpha}{2})}|\mathcal{F}_j\} \\
&= |E\{X_n + M_n 1_{(|M_n|\leq\frac{\alpha}{2})} \mid \mathcal{F}_j\} \\
&\leq |E\{X_n|\mathcal{F}_j\}| + \frac{\alpha}{2} \\
&= |Z_j| + \frac{\alpha}{2}.
\end{aligned}$$

By Doob's First Inequality (Theorem 26.1) we have

$$\begin{aligned}
P(M_n^* > \alpha) &\leq P\left(Z_n^* > \frac{\alpha}{2}\right) \\
&\leq \frac{2}{\alpha} E\{|Z_n|\} \leq \frac{2}{\alpha} E\{|X_n|\} \\
&= \frac{2}{\alpha} E\{|M_n| 1_{\{|M_n|>\frac{\alpha}{2}\}}\}.
\end{aligned}$$

By Lemma 26.1 we have

$$E\{(M_n^*)^p\} = \int_0^\infty p\lambda^{p-1}P(M_n^* > \lambda)d\lambda$$

$$\leq \int_0^\infty 2p\lambda^{p-2}E\{|M_n|1_{\{|M_n|>\frac{\lambda}{2}\}}\}d\lambda$$

and using Fubini's theorem (see Exercise 10.15):

$$= E\left\{\int_0^{2|M_n|} 2p\lambda^{p-2}d\lambda|M_n|\right\}$$

$$= \frac{2^p p}{p-1}E\{|M_n|^p\}.$$

□

Note that we showed in the proof of Theorem 26.2 that the constant $c \leq \frac{2^p p}{p-1}$. With more work one can show that $c^{\frac{1}{p}} = \frac{p}{p-1}$. Thus Theorem 26.2 could be restated as:

Theorem 26.3 (Doob's L^p Martingale Inequalities). *Let $M = (M_n)_{n\geq 0}$, be a martingale or a positive submartingale. Let $1 < p < \infty$. Then*

$$E\{(M_n^*)^p\}^{\frac{1}{p}} \leq \frac{p}{p-1}E\{|M_n|^p\}^{\frac{1}{p}},$$

or in the notation of L^p norms:

$$\|M_n^*\|_p \leq \frac{p}{p-1}\|M_n\|_p.$$

Our last inequality of this section is used to prove the Martingale Convergence Theorem of Chapter 27. We introduce Doob's notion of *upcrossings*. Let $(X_n)_{n\geq 0}$ be a submartingale, and let $a < b$. The number of upcrossings of an interval $[a, b]$ is the number of times a process crosses from below a to above b at a later time. We can express this idea nicely using stopping times. Define

$$T_0 = 0,$$

and inductively for $j \geq 0$:

$$S_{j+1} = \min\{k > T_j : X_k \leq a\}, \qquad T_{j+1} = \min\{k > S_{j+1} : X_k \geq b\}, \quad (26.1)$$

with the usual convention that the minimum of the empty set is $+\infty$; with the dual convention that the maximum of the empty set is 0, we can then define

$$U_n = \max\{j : T_j \leq n\} \qquad (26.2)$$

and U_n is the number of upcrossings of $[a, b]$ before time n.

Theorem 26.4 (Doob's Upcrossing Inequality). *Let $(X_n)_{n\geq0}$ be a submartingale, let $a < b$ and let U_n be the number of upcrossings of $[a, b]$ before time n (as defined in (26.2)). Then*

$$E\{U_n\} \leq \frac{1}{b-a} E\{(X_n - a)^+\}$$

where $(X_n - a)^+ = \max(X_n - a, 0)$.

Proof. Let $Y_n = (X_n - a)^+$. Since the function $\varphi(x) = (x - a)^+$ is convex and nondecreasing, we have by Exercise 25.6 that $(Y_n)_{n\geq0}$ is a submartingale. Since $S_{n+1} > n$, we obtain:

$$Y_n = Y_{S_1 \wedge n} + \sum_{i=1}^{n} (Y_{T_i \wedge n} - Y_{S_i \wedge n}) + \sum_{i=1}^{n} (Y_{S_{i+1} \wedge n} - Y_{T_i \wedge n}) . \qquad (26.3)$$

Each upcrossing of (X_n) between times 0 and n corresponds to an integer i such that $S_i < T_i \leq n$, with $Y_{S_i} = 0$ and $Y_{T_i} = Y_{T_i \wedge n} \geq b - a$, while $Y_{T_i \wedge n} - Y_{S_i \wedge n} \geq 0$ by construction for all i. Hence

$$\sum_{i=1}^{n} (Y_{T_i \wedge n} - Y_{S_i \wedge n}) \geq (b - a) U_n.$$

By virtue of (26.3) we get

$$(b - a) U_n \leq Y_n - Y_{S_1 \wedge n} - \sum_{i=1}^{n} (Y_{S_{i+1} \wedge n} - Y_{T_i \wedge n}),$$

and since $Y_{S_1 \wedge n} \geq 0$, we obtain

$$(b - a) U_n \leq Y_n - \sum_{i=1}^{n} (Y_{S_{i+1} \wedge n} - Y_{T_i \wedge n}).$$

Take expectations on both sides: since (Y_n) is a submartingale and the stopping times $T_i \wedge n$ and $S_{i+1} \wedge n$ are bounded (by n) and $T_i \wedge n \leq S_{i+1} \wedge n$, we have $E\{Y_{S_{i+1} \wedge n} - Y_{T_i \wedge n}\} \geq 0$ and thus

$$(b - a) E\{U_n\} \leq E\{Y_n\}.$$

$\square$

Exercises for Chapter 26

26.1 Let $Y_n \in L^2$ and suppose $\lim_{n \to \infty} E(Y_n^2) = 0$. Let $(\mathcal{F}_k)_{k \geq 0}$ be an increasing sequence of σ-algebras and let $X_k^n = E\{Y_n | \mathcal{F}_k\}$. Show that $\lim_{n \to \infty} E\{\sup_k (X_k^n)^2\} = 0$.

26.2 Let X, Y be nonnegative and satisfy

$$\alpha P(X \geq \alpha) \leq E\{Y 1_{\{X \geq \alpha\}}\},$$

for all $\alpha > 0$. Show that $E\{X^p\} \leq E\{q X^{p-1} Y\}$, where $\frac{1}{p} + \frac{1}{q} = 1; p > 1$.

26.3 Let X, Y be as in Exercise 26.2 and suppose that $\|X\|_p < \infty$ and $\|Y\|_p < \infty$. Show that $\|X\|_p \leq q\|Y\|_p$. (Hint: Use Exercise 26.2 and Hölder's inequality.)

26.4 Establish Exercise 26.3 without the assumption that $\|X\|_p < \infty$.

26.5 * Use Exercise 26.3 to prove Theorem 26.3.

27 Martingale Convergence Theorems

In Chapter 17 we studied convergence theorems, but they were all of the type that one form of convergence, plus perhaps an extra condition, implies another type of convergence. What is unusual about martingale convergence theorems is that no type of convergence is assumed – only a certain structure – yet convergence is concluded. This makes martingale convergence theorems special in analysis; the only similar situation arises in ergodic theory.

Theorem 27.1 (Martingale Convergence Theorem). *Let $(X_n)_{n \geq 1}$ be a submartingale such that $\sup_n E\{X_n^+\} < \infty$. Then $\lim_{n \to \infty} X_n = X$ exists a.s. (and is finite a.s.). Moreover, X is in L^1. [Warning: we do not assert here that X_n converges to X in L^1; this is not true in general.]*

Proof. Let U_n be the number of upcrossings of $[a, b]$ before time n, as defined in (26.2). Then U_n is non-decreasing hence $U(a, b) = \lim_{n \to \infty} U_n$ exists. By the Monotone Convergence Theorem

$$E\{U(a, b)\} = \lim_{n \to \infty} E\{U_n\}$$

$$\leq \frac{1}{b - a} \sup_n E\{(X_n - a)^+\}$$

$$\leq \frac{1}{b - a} \left(\sup_n E\{X_n^+\} + |a| \right) \leq \frac{c}{b - a} < \infty$$

for some constant c; $c < \infty$ by our hypotheses, and the first inequality above comes from Theorem 26.4 and the second one from $(x - a)^+ \leq x^+ + |a|$ for all reals a, x. Since $E\{U(a, b)\} < \infty$, we have $P\{U(a, b) < \infty\} = 1$. Then X_n upcrosses $[a, b]$ only finitely often a.s., and if we let

$$\Lambda_{a,b} = \{\limsup_{n \to \infty} X_n \geq b; \liminf_{n \to \infty} X_n \leq a\},$$

then $P(\Lambda_{a,b}) = 0$. Let $\Lambda = \bigcup_{\substack{a < b \\ a,b \in \mathbf{Q}}} \Lambda_{a,b}$ where $\mathbf{Q}$ denotes the rationals. Then $P(\Lambda) = 0$ since all rational pairs are countable; but

$$\Lambda = \{\limsup_n X_n > \liminf_n X_n\},$$

and we conclude $\lim_{n \to \infty} X_n$ exists a.s.

It is still possible that the limit is infinite however. Since X_n is a submartingale, $E\{X_n\} \geq E\{X_0\}$, hence

$$
\begin{aligned}
E\{|X_n|\} &= E\{X_n^+\} + E\{X_n^-\} \\
&= 2E\{X_n^+\} - E\{X_n\} \\
&\leq 2E\{X_n^+\} - E\{X_0\},
\end{aligned}
\tag{27.1}
$$

hence

$$
E\{\lim_n |X_n|\} \leq \liminf_{n \to \infty} E\{|X_n|\} \leq 2\sup_n E\{X_n^+\} - E\{X_0\} < \infty,
$$

by Fatou's lemma and (27.1) combined with the hypothesis that $\sup_n E\{X_n^+\} < \infty$. Thus X_n converges a.s. to a finite limit X. Note that we have also showed that $E\{|X|\} = E\{\lim_{n \to \infty} |X_n|\} < \infty$, hence X is in L^1. □

Corollary 27.1. *If X_n is a nonnegative supermartingale, or a martingale bounded above or bounded below, then $\lim_{n \to \infty} X_n = X$ exists a.s., and $X \in L^1$.*

Proof. If X_n is a nonnegative supermartingale then $(-X_n)_{n \geq 1}$ is a submartingale bounded above by 0 and we can apply Theorem 27.1.

If $(X_n)_{n \geq 1}$ is a martingale bounded below, then $X_n \geq -c$ a.s., all n, for some constant c, with $c > 0$. Let $Y_n = X_n + c$, then Y_n is a nonnegative martingale and hence a nonnegative supermartingale, and we need only to apply the first part of this corollary. If $(X_n)_{n \geq 1}$ is a martingale bounded above, then $(-X_n)_{n \geq 1}$ is a martingale bounded below and again we are done. □

Theorem 27.1 gives the a.s. convergence to a r.v. X, which is in L^1. But it does not give L^1 convergence of X_n to X. To obtain that we need a slightly stronger hypothesis, and we need to introduce the concept of uniform integrability.

Definition 27.1. *A subset $\mathcal{H}$ of L^1 is said to be a* uniformly integrable *collection of random variables if*

$$
\lim_{c \to \infty} \sup_{X \in \mathcal{H}} E\{1_{\{|X| \geq c\}} |X|\} = 0.
$$

Next we present two sufficient conditions to ensure uniform integrability.

Theorem 27.2. *Let $\mathcal{H}$ be a class of random variables*

a) *If $\sup_{X \in \mathcal{H}} E\{|X|^p\} < \infty$ for some $p > 1$, then $\mathcal{H}$ is uniformly integrable.*
b) *If there exists a r.v. Y such that $|X| \leq Y$ a.s. for all $X \in \mathcal{H}$ and $E\{Y\} < \infty$, then $\mathcal{H}$ is uniformly integrable.*

Proof. (a) Let k be a constant such that $\sup_{X \in \mathcal{H}} E\{|X|^p\} < k < \infty$. If $x \geq c > 0$, then $x^{1-p} \leq c^{1-p}$, and multiplying by x^p yields $x \leq c^{1-p}x^p$. Therefore we have

$$E\left\{|X|1_{\{|X|>c\}}\right\} \leq c^{1-p}E\left\{|X|^p 1_{\{|X|>c\}}\right\} \leq \frac{k}{c^{p-1}},$$

hence $\lim_{c \to \infty} \sup_{X \in \mathcal{H}} E\{|X|1_{\{|X|>c\}}\} \leq \lim_{c \to \infty} \frac{k}{c^{p-1}} = 0$.

(b) Since $|X| \leq Y$ a.s. for all $X \in \mathcal{H}$, we have

$$|X|1_{\{|X|>c\}} \leq Y1_{\{Y>c\}}.$$

But $\lim_{c \to \infty} Y1_{\{Y>c\}} = 0$ a.s.; thus by Lebesgue's dominated convergence theorem we have

$$\lim_{c \to \infty} \sup_{X \in \mathcal{H}} E\{|X|1_{\{|X|>c\}}\} \leq \lim_{c \to \infty} E\{Y1_{\{Y>c\}}\}$$
$$= E\{\lim_{c \to \infty} Y1_{\{Y>c\}}\} = 0.$$

$\square$

For more results on uniform integrability we recommend [15, pp. 16–21]. We next give a strengthening of Theorem 27.1 for the martingale case.

Theorem 27.3 (Martingale Convergence Theorem). *a) Let $(M_n)_{n \geq 1}$ be a martingale and suppose $(M_n)_{n \geq 1}$ is a uniformly integrable collection of random variables. Then*

$$\lim_{n \to \infty} M_n = M_\infty \text{ exists a.s.,}$$

M_∞ *is in* L^1, *and* M_n *converges to* M_∞ *in* L^1. *Moreover* $M_n = E\{M_\infty \mid \mathcal{F}_n\}$.

b) Conversely let $Y \in L^1$ *and consider the martingale* $M_n = E\{Y|\mathcal{F}_n\}$. *Then* $(M_n)_{n \geq 1}$ *is a uniformly integrable collection of r.v.'s.*

In other words, with the terminology of Definition 24.2, the martingale (M_n) is *closed* if and only if it is uniformly integrable.

Proof. a) Since $(M_n)_{n \geq 1}$ is uniformly integrable, for $\varepsilon > 0$ there exists c such that $\sup_n E\{|M_n|1_{\{|M_n| \geq c\}}\} \leq \varepsilon$. Therefore

$$E\{|M_n|\} = E\{|M_n|1_{\{|M_n| \geq c\}}\} + E\{|M_n|1_{\{|M_n| < c\}}\}$$
$$\leq \varepsilon + c.$$

Therefore $(M_n)_{n \geq 1}$ is bounded in L^1. Therefore $\sup_n E\{M_n^+\} < \infty$ and by Theorem 27.1 we have

$$\lim_{n \to \infty} M_n = M_\infty \text{ exists a.s. and } M_\infty \text{ is in } L^1.$$

To show M_n converges to M_∞ in L^1, define

$$f_c(x) = \begin{cases} c & \text{if } x > c, \\ x & \text{if } |x| \leq c, \\ -c & \text{if } x < -c. \end{cases}$$

Then f is Lipschitz. By the uniform integrability there exists c sufficiently large that for $\varepsilon > 0$ given:

$$E\{|f_c(M_n) - M_n|\} < \frac{\varepsilon}{3}, \quad \text{all } n; \tag{27.2}$$

$$E\{|f_c(M_\infty) - M_\infty|\} < \frac{\varepsilon}{3}. \tag{27.3}$$

Since $\lim M_n = M_\infty$ a.s. we have $\lim_{n \to \infty} f_c(M_n) = f_c(M_\infty)$, and so by Lebesgue's Dominated Convergence Theorem (Theorem 9.1(f)) we have for $n \geq N$, N large enough:

$$E\{|f_c(M_n) - f_c(M_\infty)|\} < \frac{\varepsilon}{3}. \tag{27.4}$$

Therefore using (27.2), (27.3), and (27.4) we have

$$E\{|M_n - M_\infty|\} < \varepsilon, \quad \text{for } n \geq N.$$

Hence $M_n \to M_\infty$ in L^1. It remains to show $E\{M_\infty \mid \mathcal{F}_n\} = M_n$. Let $\Lambda \in \mathcal{F}_m$ and $n \geq m$. Then

$$E\{M_n 1_\Lambda\} = E\{M_m 1_\Lambda\}$$

by the martingale property. However,

$$|E\{M_n 1_\Lambda\} - E\{M_\infty 1_\Lambda\}| \leq E\{|M_n - M_\infty| 1_\Lambda\}$$
$$\leq E\{|M_n - M_\infty|\}$$

which tends to 0 as n tends to ∞. Thus $E\{M_m 1_\Lambda\} = E\{M_\infty 1_\Lambda\}$ and hence $E\{M_\infty \mid \mathcal{F}_n\} = M_n$ a.s.

b) We already know that $(M_n)_{n \geq 1}$ is a martingale. If $c > 0$ we have

$$M_n 1_{\{|M_n| \geq c\}} = E\{Y 1_{\{|M_n| \geq c\}} \mid \mathcal{F}_n\},$$

because $\{|M_n| \geq c\} \in \mathcal{F}_n$. Hence for any $d > 0$ we get

$$E\{|M_n| 1_{\{|M_n| \geq c\}}\} \leq E\{|Y| 1_{\{|M_n| \geq c\}}\}$$
$$\leq E\{|Y| 1_{\{|Y| > d\}}\} + dP(|M_n| \geq c)$$
$$\leq E\{|Y| 1_{\{|Y| > d\}}\} + \frac{d}{c} E\{|M_n|\}. \tag{27.5}$$

Take $\varepsilon > 0$. We choose d such that the first term in (27.5) is smaller than $\varepsilon/2$, then c such that the second term in (27.5) is smaller than $\varepsilon/2$: thus $E\{|M_n| 1_{\{|M_n| > c\}}\} \leq \varepsilon$ for all n, and we are done. $\qquad \square$

Corollary 27.2. *Let $(\mathcal{F}_n)_{n\geq 0}$ be an increasing sequence of σ-algebras. That is, $\mathcal{F}_n$ is a sub-σ-algebra of $\mathcal{F}_{n+1}$ for each $n \geq 0$. Let $\mathcal{F}_\infty = \sigma(\cup_{n\geq 0}\mathcal{F}_n)$, the σ-algebra generated by the sequence $\mathcal{F}_n$. If $Y \in L^1(\mathcal{F}_\infty)$ then $\lim_{n\to\infty} E\{Y|\mathcal{F}_n\} = Y$, where the limit is in L^1.*

Proof. Let $M_n = E\{Y|\mathcal{F}_n\}$. Then M is a uniformly integrable martingale by part (b) of Theorem 27.3, and it converges to Y in L^1 by part (a) of Theorem 27.3. □

The martingale property is that $E\{X_m \mid \mathcal{F}_n\} = X_n$ a.s. for $m \geq n$ and it is natural to think of n, m as positive counting numbers (i.e., integers), as we did above. But we can also consider the index set $-\mathbf{N}$: the negative integers. In this case if $|m| > |n|$, but m and n are negative integers, then $m < n$. To minimize confusion, we *always* assume that m and n are nonnegative integers, and we write X_{-n}. So we start with an increasing family of σ-algebras $(\mathcal{F}_{-n})_{n\in\mathbf{N}}$, meaning here that $\mathcal{F}_{-n-1} \subset \mathcal{F}_{-n}$. Then, a *backwards martingale* is a sequence $(X_{-n})_{n\in\mathbf{N}}$ of integrable r.v., with X_{-n} being $\mathcal{F}_{-n}$-measurable and satisfying

$$E\{X_{-n}|\mathcal{F}_{-m}\} = X_{-m} \quad \text{a.s.,} \tag{27.6}$$

where $0 \leq n < m$.

Theorem 27.4 (Backwards Martingale Convergence Theorem). *Let $(X_{-n}, \mathcal{F}_{-n})_{n\in\mathbf{N}}$ be a backwards martingale, and let $\mathcal{F}_{-\infty} = \cap_{n=0}^{\infty}\mathcal{F}_{-n}$. Then the sequence (X_{-n}) converges a.s. and in L^1 to a limit X as $n \to +\infty$ (in particular X is a.s. finite and is integrable).*

Proof. Let U_{-n} be the number of upcrossings of $(-X_{-n})_{n\geq 0}$ of $[a, b]$ between time $-n$ and 0. Then U_{-n} is increasing as n increases, and let $U(a, b) = \lim_{n\to\infty} U_{-n}$, which exists. By Monotone Convergence

$$E\{U(a,b)\} = \lim_{n\to\infty} E\{U_{-n}\}$$

$$\leq \frac{1}{b-a}E\{(-X_0 - a)^+\} < \infty,$$

hence $P\{U(a,b) < \infty\} = 1$. The same upcrossing argument as in the proof of Theorem 27.1 implies $X = \lim_{n\to\infty} X_{-n}$ exists a.s.

Let $\varphi(x) = x^+ = (x \vee 0)$, which is convex and increasing and obviously $\varphi(X_{-n})$ is integrable, all n. Then Jensen's inequality (Theorem 23.9) and (27.6) imply that $X_{-n}^+ \leq E\{X_0^+|\mathcal{F}_{-n}\}$, hence $E\{X_{-n}^+\} \leq E\{X_0^+\}$. Then Fatou's lemma and the fact that $X_{-n}^+ \geq 0$ and $X_{-n}^+ \to X^+$ a.s. yield

$$E\{X^+\} \leq \liminf_n E\{X_{-n}^+\} \leq E\{X_0^+\} < \infty.$$

Henceforth $X^+ \in L^1$ and the same argument applied to the martingale $(-X_{-n})$ shows $X^- \in L^1$: thus $X \in L^1$.

It remains to prove that the convergence also takes place in L^1. To this effect, we note first that in the proof of Theorem 27.3 we have shown that if $X_{-n} \to X$ a.s., if $X \in L^1$, and if the sequence (X_{-n}) is uniformly integrable, then $X_{-n} \to X$ in L^1. Part (b) of the same theorem also shows that the family of r.v. $E\{X_0|\mathcal{G}\}$, when $\mathcal{G}$ ranges through all sub-σ-algebras of $\mathcal{F}$, is uniformly integrable. Since $X_{-n} = E\{X_0|\mathcal{F}_{-n}\}$, we readily deduce the result. $\qquad\square$

As an application of Theorem 27.4 we prove *Kolmogorov's Strong Law of Large Numbers*.

Theorem 27.5 (Strong Law of Large Numbers). *Let* $(X_n)_{n\geq 1}$ *be an i.i.d. sequence with* $E\{|X_1|\} < \infty$. *Then*

$$\lim_{n\to\infty} \frac{X_1 + \ldots + X_n}{n} = E\{X_1\} \text{ a.s..}$$

Proof. Let $S_n = X_1 + \ldots + X_n$, and $\mathcal{F}_{-n} = \sigma(S_n, S_{n+1}, S_{n+2}, \ldots)$. Then $\mathcal{F}_{-n} \subset \mathcal{F}_{-m}$ if $n \geq m$, and the process

$$M_{-n} = E\{X_1|\mathcal{F}_{-n}\}$$

is a backwards martingale. Note that $E\{M_{-n}\} = E\{X_1\}$, each n. Also note that by symmetry for $1 \leq j \leq n$:

$$E\{X_1|\mathcal{F}_{-n}\} = E\{X_j|\mathcal{F}_{-n}\} \text{ a.s.} \qquad (27.7)$$

(see Exercise 23.17). Therefore

$$M_{-n} = E\{X_1|\mathcal{F}_{-n}) = E\{X_2|\mathcal{F}_{-n}\} = \ldots = E\{X_n|\mathcal{F}_{-n}\},$$

hence

$$M_{-n} = \frac{1}{n}\sum_{j=1}^{n} E\{X_j|\mathcal{F}_{-n}\} = E\left\{\frac{S_n}{n}|\mathcal{F}_{-n}\right\} = \frac{S_n}{n} \text{ a.s.}$$

By Theorem 27.4, $\lim_{n\to\infty} E\{\frac{S_n}{n} \mid S_n, S_{n+1}, S_{n+1}, \ldots\} = X$ a.s., with $E\{X\} = E\{X_1\}$. Moreover X is measurable for the tail σ-algebra, hence by the Kolmogorov zero–one law (Theorem 10.6), we have X is constant almost surely. Thus it must equal its expectation and we are done. $\qquad\square$

Theorem 27.5, which is known as Kolmogorov's Strong Law of Large Numbers, was first published in 1933 [14], without the use of martingale theory that was developed decades later by J. L. Doob.

An application of martingale forward convergence is as follows.

Theorem 27.6 (Kolmogorov). *Let* $(Y_n)_{n\geq 1}$ *be independent random variables,* $E\{Y_n\} = 0$, *all* n, *and* $E\{Y_n^2\} < \infty$ *all* n. *Suppose* $\sum_{n=1}^{\infty} E\{Y_n^2\} < \infty$. *Let* $S_n = \sum_{j=1}^{n} Y_j$. *Then* $\lim_{n\to\infty} S_n = \sum_{j=1}^{\infty} Y_j$ *exists a.s., and it is finite a.s.*

Proof. Let $\mathcal{F}_n = \sigma(Y_1, \ldots, Y_n)$, and note that $E\{S_{n+1} - S_n \mid \mathcal{F}_n\} = E\{Y_{n+1} \mid \mathcal{F}_n\} = E\{Y_{n+1}\} = 0$, hence $(S_n)_{n \geq 1}$ is an $\mathcal{F}_n$-martingale. Note further that $\sup_n E\{S_n^+\} \leq \sup_n (E\{S_n^2\} + 1) \leq \sum_{n=1}^{\infty} E\{Y_n^2\} + 1 < \infty$. Thus the result follows from the Martingale Convergence Theorem (Theorem 27.1). □

The Martingale Convergence Theorems proved so far (Theorems 27.1 and 27.4) are strong convergence theorems: all random variables are defined on the same space and converge strongly to random variables on the same space, almost surely and in L^1. We now give a theorem for a class of martingales that do not satisfy the hypotheses of Theorem 27.1 and moreover do not have a strong convergence result. Nevertheless we can obtain a weak convergence result, where the martingale converges *in distribution* as $n \to \infty$. The limit is of course a normal distribution, and such a theorem is known as a *martingale central limit theorem*.

The result below is stated in a way similar to the Central Limit Theorem for i.i.d. variables X_n, with their partial sums S_n: Condition (i) implies that (S_n) is a martingale, but on the other hand an arbitrary martingale (S_n) is the sequence of partial sums associated with the random variables $X_n = S_n - S_{n-1}$, and these also satisfy (i).

Theorem 27.7 (Martingale Central Limit Theorem). *Let $(X_n)_{n \geq 1}$ be a sequence of random variables satisfying*

(i) $E\{X_n \mid \mathcal{F}_{n-1}\} = 0$
(ii) $E\{X_n^2 \mid \mathcal{F}_{n-1}\} = 1$
(iii) $E\{|X_n|^3 \mid \mathcal{F}_{n-1}\} \leq K < \infty$.

Let $S_n = \sum_{i=1}^n X_i$ and $S_0 = 0$. Then $\lim_{n \to \infty} \frac{1}{\sqrt{n}} S_n = Z$, where Z is $N(0, 1)$, and where the convergence is in distribution.

Proof. Convergence in distribution is of course weak convergence and we use characteristic functions to prove the theorem. For $u \in \mathbf{R}$, recall that $\varphi_X(u) = E\{e^{iuX}\}$ is the characteristic function of X. Let us define a related function by

$$\varphi_{n,j}(u) = E\left\{ e^{iu \frac{1}{\sqrt{n}} X_j} \mid \mathcal{F}_{j-1} \right\}.$$

By Taylor's theorem we have

$$e^{iu \frac{1}{\sqrt{n}} X_j} = 1 + iu \frac{1}{\sqrt{n}} X_j - \frac{u^2}{2n} X_j^2 - \frac{iu^3}{6n^{\frac{3}{2}}} \overline{X}_j^3 \tag{27.8}$$

where $\overline{X}_j$ is a (random) value in between 0 and X_j. Let us next take conditional expectations on both sides of (27.8) to get:

$$\varphi_{n,j}(u) = 1 + iu \frac{1}{\sqrt{n}} E\{X_j \mid \mathcal{F}_{j-1}\} - \frac{u^2}{2n} E\{X_j^2 \mid \mathcal{F}_{j-1}\} - \frac{iu^3}{6n^{\frac{3}{2}}} E\{\overline{X}_j^3 \mid \mathcal{F}_{j-1}\}$$

and using hypotheses (i) and (ii) we have:

$$\varphi_{n,j}(u) - 1 - \frac{u^2}{2n} = \frac{u^3}{6n^{\frac{3}{2}}} E\{\overline{X}_j^3 \mid \mathcal{F}_{j-1}\}. \tag{27.9}$$

Therefore since $|\overline{X}_j| \leq |X_j|$,
    Since $S_p = \sum_{j=1}^p X_j$, for $1 \leq p \leq n$ we have:

$$E\left\{e^{iu\frac{1}{\sqrt{n}}S_p}\right\} = E\left\{e^{iu\frac{1}{\sqrt{n}}S_{p-1}}e^{iu\frac{1}{\sqrt{n}}X_p}\right\} \tag{27.10}$$

$$= E\left\{e^{iu\frac{1}{\sqrt{n}}S_{p-1}}E\left\{e^{iu\frac{1}{\sqrt{n}}X_p} \mid \mathcal{F}_{p-1}\right\}\right\}$$

$$= E\left\{e^{iu\frac{1}{\sqrt{n}}S_{p-1}}\varphi_{n,p}(u)\right\}.$$

Using (27.10) and (27.9) we have

$$E\left\{e^{i\frac{u}{\sqrt{n}}S_p}\right\} = E\left\{e^{i\frac{u}{\sqrt{n}}S_{p-1}}\left(1 - \frac{u^2}{2n} - \frac{iu^3}{6n^{\frac{3}{2}}}\overline{X}_j^3\right)\right\}$$

and therefore

$$E\left\{e^{i\frac{u}{\sqrt{n}}S_p} - \left(1 - \frac{u^2}{2n}\right)e^{i\frac{u}{\sqrt{n}}S_{p-1}}\right\} = E\left\{e^{i\frac{u}{\sqrt{n}}S_{p-1}}\frac{iu^3}{6n^{\frac{3}{2}}}\overline{X}_j^3\right\} \tag{27.11}$$

and taking moduli of both sides of (27.11) and using hypothesis (iii) gives:

$$\left| E\left\{e^{iu\frac{1}{\sqrt{n}}S_p} - \left(1 - \frac{u^2}{2n}\right)e^{iu\frac{1}{\sqrt{n}}S_{p-1}}\right\} \right| \tag{27.12}$$

$$\leq E\left\{|e^{iu\frac{1}{\sqrt{n}}S_{p-1}}|\frac{|u|^3}{6n^{\frac{3}{2}}}E\{|X_j|^3 \mid \mathcal{F}_{j-1}\}\right\}$$

$$\leq K\frac{|u|^3}{6n^{\frac{3}{2}}}.$$

Let us fix $u \in \mathbf{R}$. Then since n tends to ∞, eventually $n \geq \frac{u^2}{2}$, and so for n large enough we have $0 \leq 1 - \frac{u^2}{2n} \leq 1$. Therefore we reduce the left side of (27.12) by multiplying by $(1 - \frac{u^2}{2n})^{n-p}$ for n large enough, to obtain

$$\left| \left(1 - \frac{u^2}{2n}\right)^{n-p} E\left\{e^{iu\frac{1}{\sqrt{n}}S_p}\right\} - \left(1 - \frac{u^2}{2n}\right)^{n-p+1} E\left\{e^{iu\frac{1}{\sqrt{n}}S_{p-1}}\right\} \right| \leq K\frac{|u|^3}{6n^{\frac{3}{2}}}. \tag{27.13}$$

Finally we use telescoping (finite) sums to observe

$$E\left\{e^{iu\frac{1}{\sqrt{n}}S_n}\right\} - \left(1 - \frac{u^2}{2n}\right)^n$$

$$= \sum_{p=1}^{n}\left(1 - \frac{u^2}{2n}\right)^{n-p} E\left\{e^{iu\frac{1}{\sqrt{n}}S_p}\right\} - \left(1 - \frac{u^2}{2n}\right)^{n-(p-1)} E\left\{e^{iu\frac{1}{\sqrt{n}}S_{p-1}}\right\}$$

and thus by the triangle inequality and (27.13) we have (always for $n \geq \frac{u^2}{2}$):

$$\left| E\left\{ e^{iu\frac{1}{\sqrt{n}}S_n} \right\} - \left(1 - \frac{u^2}{2n} \right)^n \right| \leq n \frac{K|u|^3}{6n^{\frac{3}{2}}} = K \frac{|u|^3}{6\sqrt{n}}. \tag{27.14}$$

Since the right side of (27.14) tends to 0 and

$$\lim_{n\to\infty} \left(1 - \frac{u^2}{2n} \right)^n = e^{-\frac{u^2}{2}}$$

as can be seen using L'Hôpital's rule (for example), we have that

$$\lim_{n\to\infty} E\{ e^{iu\frac{S_n}{\sqrt{n}}} \} = e^{-\frac{u^2}{2}}.$$

By Lévy's Continuity Theorem (Theorem 19.1) we have that $\frac{S_n}{\sqrt{n}}$ converges in law to Z, where the characteristic function of Z is $e^{-\frac{u^2}{2}}$; but this is the characteristic function of an $N(0,1)$ random variable (cf Example 13.5), and characteristic functions characterize distributions (Theorem 14.1), so we are done. □

Remark 27.1. If S_n is the martingale of Theorem 27.7, we know that strong martingale convergence cannot hold: indeed if we had $\lim_{n\to\infty} S_n = S$ a.s. with S in L^1, then we would have $\lim_{n\to\infty} \frac{S_n}{\sqrt{n}} = 0$ a.s., and the weak convergence of $\frac{S_n}{\sqrt{n}}$ to a normal random variable would not be possible. What makes it not possible to have the strong martingale convergence is the behavior of the conditional variances of the martingale increments X_n (hypothesis (ii) of Theorem 27.7). □

We end our treatment of martingales with an example from analysis: this example illustrates the versatile applicability of martingales; we use the martingale convergence theorem to prove a convergence result for approximation of functions.

Example 27.1. ([10]) Let f be a function in $L^p[0,1]$ for Lebesgue measure restricted to $[0,1]$. Martingale theory can provide insights into approximations of f by orthogonal polynomials.

Let us define the *Rademacher functions* on $[0,1]$ as follows. We set $R_0(x) = 1$, $0 \leq x \leq 1$. For $n \geq 1$, we set for $0 \leq x \leq 1$:

$$R_n(x) = \begin{cases} 1 & \text{if } \frac{2j-1}{2^n} \leq x < \frac{2j}{2^n}, \quad \text{some } j \text{ in } \{1,\ldots,2^n\} \\ -1 & \text{otherwise.} \end{cases}$$

We let the probability measure P be Lebesgue measure restricted to $[0,1]$, and $\mathcal{F}$ is the Borel sets of $[0,1]$. Then

$$E\{R_n\} = \int_0^1 R_n(x)dx = 0$$

and

$$\text{Var}\,(R_n) = E\{R_n^2\} = \int_0^1 R_n(x)^2 dx = 1.$$

Finally note that R_n and R_m are independent if $n \neq m$. (See Exercise 27.7.)
Next we define the *Haar functions* as follows:

$$H_0(x) = R_0(x),$$
$$H_1(x) = R_1(x).$$

For $n \geq 2$, let $n = 1 + 2 + \ldots + 2^{r-2} + \lambda = 2^{r-1} - 1 + \lambda$, where $r \geq 2$ and
$1 \leq \lambda \leq 2^{r-1}$. Then

$$H_n(x) = \begin{cases} \sqrt{2^{r-1}}R_n(x) & \text{for } \frac{2\lambda-2}{2^r} \leq x < \frac{2\lambda}{2^r}, \\ 0 & \text{otherwise.} \end{cases}$$

Next let $\mathcal{F}_n = \sigma(H_0, H_1, \ldots, H_n)$, the smallest σ-algebra making $H_0, \ldots, H_n$
all measurable. We can then check that

$$\int_\wedge H_{n+1}(x)dx = 0 \text{ if } \wedge \in \mathcal{F}_n; \tag{27.15}$$

(see Exercise 27.8.) Moreover we have

$$\int_0^1 H_n(x)dx = 0,$$
$$\int_0^1 H_n(x)^2 dx = 1.$$

We now have the following:

Theorem 27.8. *Let H_n be the Haar system on $[0,1]$ and let $f \in L^p[0,1]$ for
$p \geq 1$. Let*

$$\alpha_r = \int_0^1 H_r(x)f(x)dx,$$
$$S_n(x,f) = \sum_{r=0}^n \alpha_r H_r(x). \tag{27.16}$$

*Then $\lim_{n\to\infty} S_n(x,f) = f(x)$ a.e. Moreover if $S^*(x,f) = \sup_n |S_n(x,f)|$,
then*

$$\int_0^1 (S^*(x,f))^p dx \leq \left(\frac{p}{p-1}\right)^p \int_0^1 |f(x)|^p dx.$$

Proof. We first show that $S_n(x, f)$ is a martingale. We have

$$
\begin{aligned}
E\{S_{n+1}(x, f) \mid \mathcal{F}_n\} &= S_n(x, f) + E\{\alpha_{n+1} H_{n+1}(x) \mid \mathcal{F}_n\} \\
&= S_n(x, f) + \alpha_{n+1} E\{H_{n+1}(x) \mid \mathcal{F}_n\} \\
&= S_n(x, f)
\end{aligned}
$$

where we used (27.15). However more is true:

$$S_n(x, f) = E\{f \mid \mathcal{F}_n\}, \tag{27.17}$$

which is the key result. Indeed to prove (27.17) is where we need the coefficients α_r given in (27.16). (See Exercise 27.10.)

Next we show $S_n(x, f)$ satisfies $\sup_n E\{S_n(x, f)^+\} < \infty$, for $p > 1$ (the hypothesis for the Martingale Convergence Theorem; Theorem 27.1). We actually show more thanks to Jensen's inequality (Theorem 23.9): since $\varphi(u) = |u|^p$ is convex for $p > 1$, we have that

$$
\begin{aligned}
\int_0^1 |S_n(x, f)|^p dx &= E\{|E\{f \mid \mathcal{F}_n\}|^p\} \\
&\le E\{E\{|f|^p \mid \mathcal{F}_n\}\} \\
&= E\{|f|^p\} \\
&= \int_0^1 |f(x)|^p dx < \infty,
\end{aligned}
$$

and thus

$$
\begin{aligned}
\sup_n E\{S_n(x, f)^+\} &\le \sup_n E\{|S_n(x, f)|^p\} \\
&\le E\{|f|^p\} < \infty.
\end{aligned}
$$

We now have by Theorem 27.1 that

$$\lim_{n \to \infty} S_n(x, f) = f(x) \text{ almost everywhere,}$$

and also by Doob's L^p martingale inequalities (Theorem 26.2) we have

$$
\begin{aligned}
E\{S^*(f)^p\} &\le \left(\frac{p}{p-1}\right)^p E\{|S_n(f)|^p\} \\
&\le \left(\frac{p}{p-1}\right)^p E\{|f|^p\},
\end{aligned}
$$

or equivalently

$$\int_0^1 (S^*(x, f))^p dx \le \left(\frac{p}{p-1}\right)^p \int_0^1 |f(x)|^p dx.$$

□

We remark that results similar to Theorem 27.8 above hold for classical Fourier series, although they are harder to prove.

Exercises for Chapter 27

27.1 (A martingale proof of Kolmogorov's zero–one law.) Let X_n be independent random variables and let $\mathcal{C}_\infty$ be the corresponding tail σ–algebra (as defined in Theorem 10.6). Let $C \in \mathcal{C}_\infty$. Show that $E\{1_C|\mathcal{F}_n\} = P(C)$, all n, where $\mathcal{F}_n = \sigma(X_j; 0 \leq j \leq n)$. Show further $\lim_{n\to\infty} E\{1_C|\mathcal{F}_n\} = 1_C$ a.s. and deduce that $P(C) = 0$ or 1.

27.2 A martingale $X = (X_n)_{n\geq 0}$ is *bounded in* L^2 if $\sup_n E\{X_n^2\} < \infty$. Let X be a martingale with X_n in L^2, each n. Show that X is bounded in L^2 if and only if

$$\sum_{n=1}^{\infty} E\{(X_n - X_{n-1})^2\} < \infty.$$

(*Hint:* Recall Exercise 24.12.)

27.3 Let X be a martingale that is bounded in L^2; show that $\sup_n E\{|X_n|\} < \infty$, and conclude that $\lim_{n\to\infty} X_n = Y$ a.s., with $E\{|Y|\} < \infty$.

27.4 * Let X be a martingale bounded in L^2. Show that $\lim_{n\to\infty} X_n = X$ a.s. and in L^2. That is, show that $\lim_{n\to\infty} E\{(X_n - X)^2\} = 0$.

27.5 (*Random Signs*) Let $(X_n)_{n\geq 1}$ be i.i.d. with $P(X_n = 1) = P(X_n = -1) = \frac{1}{2}$. Let $(\alpha_n)_{n\geq 1}$ be a sequence of real numbers. Show that $\sum_{n=1}^{\infty} \alpha_n X_n$ is a.s. convergent if $\sum_{n=1}^{\infty} \alpha_n^2 < \infty$.

27.6 Let $X_1, X_2, \ldots$ be i.i.d. nonnegative random variables with $E\{X_1\} = 1$. Let $R_n = \prod_{i=1}^{n} X_i$, and show that R_n is a martingale for the σ-algebras $\mathcal{F}_n = \sigma(X_1, \ldots, X_n)$.

27.7 Show that if $n \neq m$, then the Rademacher functions R_n and R_m are independent for $P = \lambda$ Lebesgue measure restricted to $[0, 1]$.

27.8 Let H_n be the Haar functions, and suppose $\wedge \in \mathcal{F}_n = \sigma(H_0, H_1, \ldots, H_n)$. Show that

$$\int_\wedge H_{n+1}(x)dx = 0.$$

27.9 Let f be in $L^p[0, 1]$. Let $S_n(x, f)$ be as defined in (27.16) and show that $E\{f \mid \mathcal{F}_n\} = S_n(x, f)$. (*Hint:* Show that

$$\int_\wedge f(x)dx = \int_\wedge S_n(x, f)dx \text{ for } \wedge \in \mathcal{F}_n$$

by using that the Haar functions are an orthonormal system; that is,

$$\int_0^1 H_n(x)H_m(x)dx = 0 \text{ if } n \neq m \text{ and } \int_0^1 H_n(x)^2 dx = 1.)$$

27.10 Use Martingale Convergence to prove the following $0-1$ law. Let $(\mathcal{F}_n)$ be an increasing sequence of σ-algebras and $\mathcal{G}_n$ a decreasing sequence of σ-algebras, with $\mathcal{G}_1 \subset \sigma(\cup_{n=1}^{\infty}\mathcal{F}_n)$. Suppose that $\mathcal{F}_n$ and $\mathcal{G}_n$ are independent for each n. Show that if $\wedge \in \cap_{n=1}^{\infty}\mathcal{G}_n$, then $P(\wedge) = 0$ or 1.

27.11 Let $\mathcal{H}$ be a subset of L^1. Let G be defined on $[0, \infty)$ and suppose G is positive, increasing, and

$$\lim_{t\to\infty} \frac{G(t)}{t} = \infty.$$

Suppose further that $\sup_{X\in\mathcal{H}} E\{G(X)\} < \infty$. Show that $\mathcal{H}$ is uniformly integrable. (This extends Theorem 27.2(a).)

28 The Radon-Nikodym Theorem

Let $(\Omega, \mathcal{F}, P)$ be a probability space. Suppose a random variable $X \geq 0$ a.s. has the property $E\{X\} = 1$. Then if we define a set function Q on $\mathcal{F}$ by

$$Q(\wedge) = E\{1_\wedge X\} \tag{28.1}$$

then it is easy to see that Q defines a new probability (see Exercise 9.5). Indeed

$$Q(\Omega) = E\{1_\Omega X\} = E\{X\} = 1$$

and if $A_1, A_2, A_3, \ldots$ are disjoint in $\mathcal{F}$ then

$$Q\left(\bigcup_{i=1}^\infty A_i\right) = E\{1_{\cup_{i=1}^\infty A_i} X\}$$

$$= E\left\{\sum_{i=1}^\infty 1_{A_i} X\right\}$$

$$= \sum_{i=1}^\infty E\{1_{A_i} X\}$$

$$= \sum_{i=1}^\infty Q(A_i)$$

and we have countable additivity. The interchange of the expectation and the summation is justified by the Monotone Convergence Theorem (Theorem 9.1(d)).

Let us consider two properties enjoyed by Q:

(i) If $P(\wedge) = 0$ then $Q(\wedge) = 0$. This is true since $Q(\wedge) = E\{1_\wedge X\}$, and then $1_\wedge$ is a.s. 0, and hence $1_\wedge X = 0$ a.s.
(ii) For every $\varepsilon > 0$ there exists $\delta > 0$ such that if $\wedge \in \mathcal{F}$ and $P(\wedge) < \delta$, then $Q(\wedge) < \varepsilon$.

Indeed property (ii) follows from Property (i) in general. We state it formally.

Theorem 28.1. *Let P, Q be two probabilities such that $P(\wedge) = 0$ implies $Q(\wedge) = 0$ for all $\wedge \in \mathcal{F}$. Then for each $\varepsilon > 0$ there exists $\delta > 0$ such that if $\wedge \in \mathcal{F}$ and $P(\wedge) < \delta$, then $Q(\wedge) < \varepsilon$.*

Proof. Suppose the result were not true. Then there would be a sequence $\wedge_n \in \mathcal{F}$ with $P(\wedge_n) < \frac{1}{2^n}$ (for example) and $Q(\wedge_n) \geq \varepsilon$, all n, for some $\varepsilon > 0$. Set $\wedge = \limsup_{n \to \infty} \wedge_n$. By Borel-Cantelli Lemma (Theorem 10.5) we have $P(\wedge) = 0$. Fatou's lemma has a symmetric version for limsups, which we established in passing during the proof of Theorem 9.1(f); this gives

$$Q(\wedge) \geq \limsup_{n \to \infty} Q(\wedge_n) \geq \varepsilon,$$

and we obtain a contradiction. □

It is worth noting that conditions (i) and (ii) are actually *equivalent*. Indeed we showed (i) implies (ii) in Theorem 28.1; that (ii) implies (i) is simple: suppose we have (ii) and $P(\wedge) = 0$. Then for any $\varepsilon > 0$, $P(\wedge) < \delta$ and so $P(\wedge) < \varepsilon$. Since ε was arbitrary we must have $Q(\wedge) = 0$.

Definition 28.1. *Let P, Q be two finite measures. We say Q is absolutely continuous with respect to P if whenever $P(\wedge) = 0$ for $\wedge \in \mathcal{F}$, then $Q(\wedge) = 0$. We denote this $Q \ll P$.*

Examples: We have seen that for any r.v. $X \geq 0$ with $E\{X\} = 1$, we have $Q(\wedge) = E\{1_\wedge X\}$ gives a probability measure with $Q \ll P$.

A naturally occurring example is $Q(\wedge) = P(\wedge \mid A)$, where $P(A) > 0$. It is trivial to check that $P(\wedge) = 0$ implies $Q(\wedge) = 0$. Note that this example is also of the form $Q(\wedge) = E\{1_\wedge X\}$, where $X = \frac{1}{P(A)} 1_A$.

The Radon–Nikodym theorem characterizes all absolutely continuous probabilities. Indeed we see that if $Q \ll P$, then Q must be of the form (28.1). Thus our original class of examples is all that there is. We first state a simplified version of the theorem, for separable σ-fields. Our proof follows that of P. A. Meyer [15].

Definition 28.2. *A sub σ-algebra $\mathcal{G}$ of $\mathcal{F}$ is separable if $\mathcal{G} = \sigma(A_1, \ldots, A_n, \ldots)$, with $A_i \in \mathcal{F}$, all i. That is, $\mathcal{G}$ is generated by a countable sequence of events.*

Theorem 28.2 (Radon-Nikodym). *Let $(\Omega, \mathcal{F}, P)$ be a probability space with a separable σ-algebra $\mathcal{F}$. If Q is a finite measure on $\mathcal{F}$ and if $P(\wedge) = 0$ implies $Q(\wedge) = 0$ for any such $\wedge \in \mathcal{F}$, then there exists a unique integrable positive random variable X such that*

$$Q(\wedge) = E\{1_\wedge X\}.$$

We write $X = \frac{dQ}{dP}$. Further X is unique almost surely: that is if X' satisfies the same properties, then $X' = X$ P-a.s.

Proof. Since the result is obvious when $Q = 0$, we can indeed assume that $Q(\Omega) > 0$. Then we can normalize Q by taking $\tilde{Q} = \frac{1}{Q(\Omega)} Q$, so we assume without loss that Q is a probability measure. Let $A_1, A_2, \ldots, A_n$ be a countable enumeration of sets in $\mathcal{F}$ such that $\mathcal{F} = \sigma(A_1, A_2, \ldots, A_n, \ldots)$. We define an increasing family of σ-algebras $(\mathcal{F}_n)_{n \geq 1}$ by

$$\mathcal{F}_n = \sigma(A_1, \ldots, A_n).$$

There then exists a finite partition of Ω into $\mathcal{F}_n$-measurable sets $A_{n,1}, A_{n,2}, \ldots, A_{n,k_n}$ such that each element of $\mathcal{F}_n$ is the (finite) union of some of these events. Such events are called "*atoms*". We define

$$X_n(\omega) = \sum_{i=1}^{k_n} \frac{Q(A_{n,i})}{P(A_{n,i})} 1_{A_{n,i}}(\omega) \tag{28.2}$$

with the convention that $\frac{0}{0} = 0$ (since $Q \ll P$ the numerator is 0 whenever the denominator is 0 above). We wish to show the process $(X_n)_{n \geq 1}$ is in fact a martingale. Observe first that X_n is $\mathcal{F}_n$-measurable. Next, let $m \leq n$. Then exactly as in the proof of Theorem 24.6, in order to get $E\{X_n | \mathcal{F}_m\} = X_m$ it is enough to prove that for every $\wedge \in \mathcal{F}_m$ we have

$$\int_\wedge X_n dP = \int_\wedge X_m dP. \tag{28.3}$$

We can write

$$\int_\wedge X_n dP = \int_\wedge \sum_{i=1}^{k_n} \frac{Q(A_{n,i})}{P(A_{n,i})} 1_{A_{n,i}} dP$$

$$= \int \sum_{i=1}^{k_n} \frac{Q(A_{n,i})}{P(A_{n,i})} 1_{A_{n,i} \cap \wedge} dP$$

$$= \sum_{i=1}^{k_n} \frac{Q(A_{n,i})}{P(A_{n,i})} P(A_{n,i} \cap \wedge).$$

Now, since $\wedge \in \mathcal{F}_n$, the set $\wedge$ can be written as the union of some of the (disjoint) partition sets $A_{n,i}$, that is $\wedge = \cup_{i \in I} A_{n,i}$ for a subset $I \subset \{1, \ldots, k_n\}$. Therefore $\wedge \cap A_{n,i} = A_{n,i}$ if $i \in I$ and $\wedge \cap A_{n,i} = \phi$ otherwise, and we now obtain

$$\int_\wedge X_n dP = \sum_{i \in I} \frac{Q(A_{n,i})}{P(A_{n,i})} P(A_{n,i} \cap \wedge)$$

$$= \sum_{i \in I} Q(A_{n,i}) = Q(\wedge)$$

where we have used again the fact that $Q(A_{n,i}) = 0$ whenever $P(A_{n,i}) = 0$. Since $\wedge \in \mathcal{F}_m$ we get similarly $\int_\wedge X_m dP = Q(\wedge)$. Hence (28.1) holds, and further if we take $\wedge = \Omega$ then we get $\int X_n dP = Q(\Omega) = 1 < \infty$, so X_n is P-integrable. Therefore $(X_n)_{n \geq 1}$ is a martingale.

We also have that the martingale (X_n) is uniformly integrable. Indeed, we have

$$\int_{\{X_n \geq c\}} X_n dP = Q(X_n > c);$$

by Markov's inequality

$$P(X_n \geq c) \leq \frac{E\{X_n\}}{c} = \frac{1}{c}.$$

Let $\varepsilon > 0$, and let δ be associated with ε as in Theorem 28.1 (since $Q \ll P$ by hypothesis). If $c > 1/\delta$ then we have $P(X_n \geq c) < \delta$, hence $Q(X_n \geq c) \leq \varepsilon$, hence $\int_{\{X_n \geq c\}} X_n dP \leq \varepsilon$: therefore the sequence (X_n) is uniformly integrable, and by our second Martingale Convergence Theorem (Theorem 27.3) we have that there exists a r.v. X in L^1 such that $\lim_{n \to \infty} X_n = X$ a.s. and in L^1 and moreover

$$E\{X \mid \mathcal{F}_n\} = X_n.$$

Let now $\wedge \in \mathcal{F}$, and define $R(\wedge) = E\{1_\wedge X\}$. Then R agrees with Q on each $\mathcal{F}_n$, since if $\wedge \in \mathcal{F}_n$, $R(\wedge) = E\{1_\wedge X\} = E\{1_\wedge X_n\} = Q(\wedge)$. The Monotone Class Theorem (6.3) now implies that $R = Q$, since $\mathcal{F} = \sigma(\mathcal{F}_n; n \geq 1)$.   □

Remark 28.1. We can use Theorem 28.2 to prove a more general Radon–Nikodym theorem, without the separability hypothesis. For a proof of Theorem 28.3 below, see [25, pp.147–149].

Theorem 28.3 (Radon-Nikodym). *Let P be a probability on $(\Omega, \mathcal{F})$ and let Q be a finite measure on $(\Omega, \mathcal{F})$. If $Q \ll P$ then there exists a nonnegative r.v. X such that $Q(\wedge) = E\{1_\wedge X\}$ for all $\wedge \in \mathcal{F}$. Moreover X is P-unique a.s. We write $X = \frac{dQ}{dP}$.*

The Radon–Nikodym theorem is directly related to conditional expectation. Suppose given $(\Omega, \mathcal{F}, P)$ and let $\mathcal{G}$ be a sub σ-algebra of $\mathcal{F}$. Then for any nonnegative r.v. X with $E\{X\} < \infty$, $Q(\wedge) = E\{X1_\wedge\}$ for $\wedge$ in $\mathcal{G}$ defines a finite measure on $(\Omega, \mathcal{G})$, and $P(\wedge) = 0$ implies $Q(\wedge) = 0$. Thus $\frac{dQ}{dP}$ exists *on the space* $(\Omega, \mathcal{G})$, and we define $Y = \frac{dQ}{dP}$; then Y is $\mathcal{G}$-measurable. Note further that if $\wedge \in \mathcal{G}$, then

$$E\{Y1_\wedge\} = Q(\wedge) = E\{X1_\wedge\}.$$

Thus Y is a version of $E\{X \mid \mathcal{G}\}$. In fact, it is possible to prove the Radon–Nikodym Theorem with a purely measure-theoretic proof, not using martingales. Then one can *define* the conditional expectation as above: this is an alternative way for constructing conditional expectation, which does not use Hilbert space theory.

Finally note that if P is a probability on **R** having a density f, and since $P(A) = \int_A f(x)dx$, then P is absolutely continuous with respect to Lebesgue measure m on **R** (here m is a σ-finite measure, but the Radon-Nikodym Theorem "works" also in this case), and we sometimes write $f = \frac{dP}{dm}$.

Exercises for Chapter 28

28.1 Suppose Q and P are finite measures, with $Q \ll P$ and $P \ll Q$. We say that Q is *equivalent* to P, and we write $Q \sim P$. Show that $X = \frac{dQ}{dP}$ satisfies $X > 0$ almost everywhere (dP). That is, $P(X \leq 0) = 0$.

28.2 Suppose $Q \sim P$. Let $X = \frac{dQ}{dP}$. Show that $\frac{1}{X} = \frac{dP}{dQ}$ (see Exercise 9.8).

28.3 Let μ be a measure such that $\mu = \sum_{n=1}^{\infty} \alpha_n P_n$, for P_n probability measures and $\alpha_n > 0$, all n. Suppose $Q_n \ll P_n$ each n, and that $\nu = \sum_{n=1}^{\infty} \beta_n Q_n$ and $\beta_n \geq 0$, all n. Show that $\mu(\wedge) = 0$ implies $\nu(\wedge) = 0$.

28.4 Let P, Q be two probabilities and let $R = \frac{P+Q}{2}$. Show that $P \ll R$.

28.5 Suppose $Q \sim P$. Give an example of a P martingale which is not a martingale for Q. Also give an example of a process which is a martingale for both P and Q simultaneously.

References

1. R. Bass (1995), *Probabilistic Techniques in Analysis*; Springer-Verlag; New York.
2. H. Bauer (1996), *Probability Theory*; Walter de Gruyter; Berlin.
3. J. Bernoulli (1713), *Ars Conjectandi*; Thurnisiorum; Basel (Switzerland).
4. G. Cardano (1961), *Liber de ludo aleae; The Book on Games of Chance*; Sidney Gould (Translator); Holt, Rinehart and Winston.
5. G. Casella and R. L. Berger (1990), *Statistical Inference*; Wadsworth; Belmont, CA.
6. A. De Moivre (1718), *The Doctrine of Chances; or, A Method of Calculating the Probability of Events in Play*; W. Pearson; London. Also in 1756, *The Doctrine of Chances (Third Edition)*, reprinted in 1967: Chelsea, New York.
7. J. Doob (1994), *Measure Theory*; Springer-Verlag; New York.
8. R. Durrett (1991), *Probability: Theory and Examples*; Wadsworth and Brooks/Cole; Belmont, CA.
9. W. Feller (1971), *An Introduction to Probability Theory and Its Applications* (Volume II); John Wiley; New York.
10. A. Garsia (1970), *Topics in Almost Everywhere Convergence*; Markham; Chicago.
11. A. Gut (1995), *An Intermediate Course in Probability*; Springer-Verlag; New York.
12. N. B. Haaser and J. A. Sullivan (1991), *Real Analysis*; Dover; New York.
13. C. Huygens (1657); See *Oeuvres Complètes de Christiaan Huygens*, (with a French translation (1920)), The Hague: Nijhoff.
14. A. N. Kolmogorov (1933), *Grundbegriffe der Wahrscheinlichkeitrechnung*. English translation: *Foundations of the Theory of Probability*, (1950), Nathan Morrison translator; Chelsea; New York.
15. P. A. Meyer (1966), *Probability and Potentials*; Blaisdell; Waltham, MA (USA).
16. J. Neveu (1975), *Mathematical foundations of the calculus of probabilities*; Holden–Day; San Francisco.
17. D. Pollard (1984), *Convergence of Stochastic Processes*; Springer-Verlag; New York.
18. M. H. Protter and P. Protter (1988), *Calculus with Analytic Geometry* (Fourth Edition); Jones and Bartlett; Boston.
19. M. Sharpe (1988), *General Theory of Markov Processes*; Academic Press; New York.
20. G. F. Simmons (1963), *Introduction to Topology and Modern Analysis*; McGraw-Hill; New York.
21. S. M. Stigler (1986), *The History of Statistics: The measurement of uncertainty before 1900*; Harvard University Press; Cambridge, MA.
22. D. W. Stroock (1990), *A Concise Introduction to the Theory of Integration*; World Scientific; Singapore.

23. S. J. Taylor (1973), *Introduction to Measure and Integration*; Cambridge University Press; Cambridge (U.K.).
24. P. van Beek (1972), *An application of Fourier methods to the problem of sharpening the Berry-Esseen inequality, Z. Wahrscheinlichkeitstheorie und verw. Gebiete* **23**, 183–196.
25. D. Williams (1991), *Probability with Martingales*; Cambridge; Cambridge, UK.

Index